SHORT-RANGE OPTICAL WIRELESS

SHORT-RANGE OPTICAL WIRELESS

THEORY AND APPLICATIONS

Mohsen Kavehrad
The Pennsylvania State University
University Park, PA, USA

M. I. Sakib Chowdhury
The Pennsylvania State University
University Park, PA, USA

Zhou Zhou
The Pennsylvania State University
University Park, PA, USA

Library of Congress Cataloging-in-Publication data applied for

ISBN: 9781118887707

A catalogue record for this book is available from the British Library.

Set in 10/12pt Times by SPi Global, Pondicherry, India
Printed and bound in Singapore by Markono Print Media Pte Ltd

1   2016

Contents

Preface

This book provides a detailed overview of physical layer optical wireless communications. The basic pedagogic methodology is to include fully detailed derivations from the basic principles. The text is intended to provide enough principle materials to guide the novice student, while at the same time to have plenty of detailed materials to satisfy graduate students inclined to pursue research in the area. The book is intended to stress the principles of optical wireless communications that are useful for a wide array of applications that these techniques are devised for. It is also intended to serve as a possible textbook and reference for graduate students and a reference for practicing engineers.

Organization of the Book

In Chapter 1, we offer an introduction to optical wireless field. This covers a broad array of issues like potential of solving complicated communications problems considering a shortage of radio frequency spectrum suitable for mobile applications and suggestion of moving some of the less mobile applications to the optical frequencies range of infrared and visible light, radio frequency interference, and necessity of transmission at very high data rates, etc. by optical wireless systems. Optical wireless links can establish communications channels even millions of miles apart, as evidenced by the usage of optical links in space exploratory missions. For shorter terrestrial distances, optical wireless links in outdoor free space are a good choice for establishing pointed links few miles apart.

In general, the optical wireless communications area of research did not receive much attention for several years, except in some military applications for the security that it offers. Yet the largest number of wireless devices ever sold, namely TV remote controls use wireless infrared light in order to function. The advantages of using optical radiation over RF include the following:

- Virtually unlimited bandwidth with over 540 THz for wavelengths in the range of 200–1550 nm. This band is unregulated and available for immediate utilization.

- Use of baseband digital technology.
- A small receiver (photodetector) area provides spatial diversity that eliminates multipath fading in intensity modulation with direct detection links. Multipath fading degrades the performance of an unprotected RF link.
- Light is absorbed by dark surfaces, diffusely reflected by light-colored objects and directionally reflected from shiny surfaces. It does not penetrate opaque objects. This provides spatial confinement that prevents interference between adjacent cells operating in environments separated by opaque dividers.
- Spatial confinement of optical signals allows for secure data exchange without the fear of an external intruder listening in. This provides physical layer security which is the safest type.
- No electromagnetic interference with other devices, making it very suitable for environments employing interference-sensitive devices, such as hospitals, airports and factories, power plants, and military and national security buildings.

Visible light (VL) applications are emerging technology areas that utilize the high-speed switching properties of VL LEDs for wireless data applications with data rates higher than conventional 802.11 wireless networks and have additional benefits of the following:

- Sustainable solution for the current **Spectrum Crunch.**
- Energy efficiency in luminous efficacy—LEDs are far more efficient than incandescent and far more flexible than compact fluorescent lights (CFLs).

As LEDs increasingly displace incandescent lighting over the next few years, general applications of VL technology are expected to include wireless Internet access, vehicle-to-vehicle communications, broadcast from LED signage, machine-to-machine communications, positioning systems, and navigation. Furthermore, since smart LEDs have IP addresses, each will add a node to the Internet. Hence, *the most compelling story of how Internet of Light will transform our world is the one still being written: the future of lighting/communications/sensing and the birth of a new enterprise lighting network.*

As the next step to delve further into optical wireless communications systems, we discuss some fundamentals of this technology in Chapter 2. We point out the differences between radio-frequency-based and optical wireless communications systems and provide some details on optical transmitters and receivers.

In Chapter 3, we aim to establish a proper channel modeling method to ensure that modeling can be done accurately and fast. Channel modeling helps us to understand effects of intersymbol interference (ISI) at high data rates which is caused by multiple reflections of optical signals from the walls within a room.

Our goal in Chapter 4 is to analyze various channel properties using the models obtained from channel modeling techniques. We discuss different topics related to indoor optical wireless channels such as average delay spread and path loss. We have shown effects of additional room furniture on these parameters compared to an empty room.

Chapter 5 consists of several fundamental results on VL communications (VLC) utilizing multiple sources. Source layout is one of the most important factors that affect overlapping of light footprints and thus produce ISI. We explore VLC performance in conventional household layouts and investigate the impact of these layouts to VLC.

Orthogonal frequency division multiplexing (OFDM) is a good candidate for VLC as it offers robustness against multipath dispersion caused by diffuse indoor optical wireless (OW) channel. But OFDM systems contain certain disadvantages like high peak-to-average power ratio (PAPR) and an additional requirement in optical domain that the modulating signal has to be unipolar. In Chapter 6, we develop some techniques to reduce high PAPR in OFDM-based OW systems since the non-linear characteristics of LED transmitters can severely affect system performance. We then analyze performance of various OFDM-based OW schemes in multipath diffuse indoor wireless channel.

In Chapter 7, multiple-input and multiple-output (MIMO) techniques are included as they provide either reliability improvement or bandwidth efficiency increase.

In Chapter 8, based on these investigations, we further explore VLC performance in real applications, such as aircraft cabin wireless communications.

In Chapter 9, we discuss about multi-spot diffusing configuration where multiple spots on the ceiling of a room are created with the help of a holographic diffuser. We also discuss about angle-diversity receivers which are set up in such a way that several photodiodes are arranged at different angles relative to one another and thus face different directions, or can be constructed using holographic mirrors.

In Chapter 10, we investigate several fundamental research topics of indoor positioning and navigation systems based on VLC technology. Despite the fact that indoor positioning has become an attractive research topic within the past two decades, no satisfying solution has been found with consideration of both accuracy and system complexity. Recently, research on VLCs offers new opportunities on realizing accurate indoor positioning with relatively simple system configurations. We compare and discuss several positioning algorithms and describe an asynchronous positioning system.

Acknowledgments

I am grateful to my past and present doctoral students, in particular, both my coauthors, who have contributed to this book through their thesis research.

Mohsen Kavehrad
The Pennsylvania State University,
University Park, PA, USA

1

Introduction

1.1 Motivation

Optical wireless communications (OWC) has become an increasingly important research area. The potential of solving complicated communications problems, such as the shortage of radio frequency (RF) spectrum, interference, and the necessity of transmission at very high data rates by optical wireless systems has seen vast improvement. Optical wireless links can establish communications channels even millions of miles apart, as evidenced by the use of optical links in space exploratory missions by NASA [1]. For shorter terrestrial distances, optical wireless links in outdoor free space are a good choice for establishing pointed links a couple of miles apart. On a much smaller scale, the existence of millions of remote controls that operate using infrared light-emitting-diodes (LEDs) is a proof of the usefulness of optical wireless systems.

Apart from the various applications of OWC that are currently in use, probably the main motivating factor to focus on this area is the possibility of mitigating the increasing spectrum shortage issue. As consumption of high data rate multimedia materials is increasing day by day and the use of handheld devices is becoming more and more widespread, the precious RF spectrum range of about 1.9 GHz that is used for mobility is getting scarcer [2]. Users are encouraged to shift to the Wi-Fi bands instead of the bands used for cellular services in order to alleviate this increasing load of high data rate applications. However, there are places where even Wi-Fi bands do not operate as expected or are found to be so congested that their use becomes next to impossible, for example, heavily crowded conference halls. Also, supported data rates of Wi-Fi as well as cellular data services should be considered in this discussion. Though IEEE 802.11ac and IEEE 802.11ad standards are supposed to support high bit rates, they are not yet widespread, and so the cost issue is involved. LTE and LTE-Advanced standards are also supposed to support high bit rates, but they use the same precious cellular

Short-Range Optical Wireless: Theory and Applications, First Edition. Mohsen Kavehrad,
M. I. Sakib Chowdhury and Zhou Zhou.

spectrum band and thus due to congestion cannot provide satisfactory performance. Hence, the pursuit of and research on alternatives to these radio frequency-based solutions such as optical wireless-based systems and technologies are greatly desirable [3, 4].

OWC can be both indoors and outdoors and are usually broadly divided into two categories based on the type of optical source employed. Two types of optical sources—LEDs and lasers—are currently in use as transmitters of optical links. The difference between these two sources lies in their supported bandwidth: where LEDs have a much lower electrical bandwidth than lasers, and hence if very high data rate transmission in the range of Gbps is required, lasers are the popular choice. Also, lasers emit monochromatic light signals, that is, light signals that have only one wavelength in it, whereas LEDs have a very broad spectral line-width. LED-based communications mainly involve visible light communications (VLC) using white LEDs (WLEDs), and lasers are used only as very high-speed infrared sources. Hence, these two types of optical sources have different application scenarios. In this book, we will cover different types of applications where both LEDs and lasers are used.

The energy-saving aspect of WLEDs is probably one of the most important benefits that can be obtained using VLC. Lighting is a major source of electric energy consumption. It is estimated that one-third of the global consumption of electricity is spent for lighting purposes; therefore, development of more efficient lighting sources is important. This acknowledgment of concerns about significant consumption has generated significant activity toward the development of solid-state sources, to replace incandescent and fluorescent lights. Fluorescent lamps contain environmental pollutants, thus their elimination will remove a significant source of environmental pollution and more specifically, their replacement with highly efficient LEDs generating "white light" will reduce energy consumption. It is fortunate that WLEDs are already commercially available. WLEDs require roughly 20 times less power compared to conventional light sources, even 5 times less power compared to fluorescent bulbs that consume less energy. An entire rural village can be lit with less energy than that used by a single conventional 100 W light bulb. Switching to solid-state lighting would reduce global electricity use by 50% and reduce power consumption by 760 GW in the United States alone over a 20-year period. To get a clear picture of the positive impact the use of WLEDs will have, some concrete estimates can be provided. If all existing bulbs were replaced by WLED sources, within 10 years we will have the following benefits: energy savings of 1.9×10^{20} J, US\$1.83 trillion financial savings, 10.68 GT reduction of carbon dioxide emissions, and 962 million barrels less consumption of crude oil [4].

The field of photonics starts with the efficient generation of light. The generation of efficient yet highly controllable light can indeed be accomplished using LEDs. Using a WLED instead of conventional lighting means the size, cost, and energy consumption will decrease considerably, as optical devices are smaller and simpler than electrical devices. WLEDs are semiconductor devices. About 13 000 LEDs can be formed on a substrate, which can be about 0.25×0.25 units in size. WLEDs use 5% of the energy of a regular incandescent bulb. An entire rural village can be lit with less energy than that used by a single conventional 100 W light bulb. By replacing the conventional lights with WLEDs and by using them for both data transmission and lighting, large amounts of energy can be saved. Undoubtedly, white light emitting solid-state devices will be the lighting sources of the twenty-first century. About 10–15 years ago, researchers came to the realization that WLED devices, in addition to being very fit for lighting the surrounding space, could also be used for wireless communications purposes. The advantages of such technology applications are many. It belongs to the "green technologies" category when used for lighting purposes, becoming even more environmentally friendly when it supports communication functionality compared to RF alternatives. Also, LEDs and photodetectors tend to be considerably cheaper compared to RF counterparts. OWC allows easy bandwidth reuse and improves security, as light is confined within the room it illuminates. It does not generate RF contamination, nor is it impacted by RF interference. Thus, replacing RF devices with devices using white light for communications (at least for indoor environments) will reduce interference in the RF bands. It should be pointed out that while the consumer market and the product developers will benefit from this, the technology can also make a major breakthrough in cases where RF radiation is of great concern, as in the case of hospitals, schools, airplanes, and mines. RF interference has caused accidental triggering of explosions when using remote detonator devices. Federal regulation places 1 W as the maximum acceptable RF power within mines using remotely triggered detonators. Also, baby monitoring RF signals have interfered with landing instructions of planes approaching airport runways.

1.1.1   Spectrum Scarcity Issues and Optical Wireless Communications as a Solution

Let us delve a bit deeper into the RF spectrum scarcity problem that we mentioned earlier and how OWC using either LEDs or lasers can help in this regard.

With the increasing popularity of multimedia services supplied over the RF networks and services such as web browsing, audio and video on demand, it is for sure only a matter of time before users will face extreme congestion while trying to connect to avail themselves of these aforementioned services. Advancements in displays, battery technology, and processing power have made it possible for users to afford and carry around smart phones and tablets. As we are entering a new era of always on connectivity, the expectation from users for not only ubiquitous but also seamless voice and video services presents a significant challenge for today's telecommunications systems. The prospects for the delivery of such multimedia services to these users are crucially dependent on the development of low-cost physical layer delivery mechanisms.

According to market research published by Cisco Systems, Inc. [5], the largest manufacturer of networking equipment, mobile data consumption is going to explode in the next 5 years, largely due to the proliferation of mobile video and mobile web applications. Cisco market research includes the Visual Networking Index (VNI). The VNI research predicts mobile data use to expand from 2.5 to 24.3 EB monthly. This is an increase of a factor of 10 in 5 years, or about 57% cumulative annual growth rate (CAGR). This is an enormous growth in mobile data, a very large portion of which is growth due to the proliferation of mobile video (66%). Much of this mobile data growth (about 70%) will be consumed by laptops and other mobile ready portables such as pico-projectors, wireless reading devices, digital photo frames, and smart phones. These mobile devices can generally be thought of as in-building networked devices that are used to share information (video) within a classroom, conference, or meeting room. The report predicts that a greater amount of traffic will migrate from fixed to mobile networks.

In the past few years, we have witnessed rapid growth in technologies producing low-cost communications devices, using the RF license-free bands: ISM (2.4–2.4835 GHz), UNII

(5.15–5.25 and 5.35–5.825 GHz). As technology advances, the service capability of such devices will strengthen. However, uncontrolled deployment of devices using the same spectrum allocation can generate interference beyond the level that these systems can afford, thus leading to service quality deterioration. The IEEE 802.15.2 working group was formed to address this growing problem; however, without controlling the number of devices operating within certain areas, the problem cannot be solved, unless more bandwidth becomes available. The 57–64 GHz band has been added to license free bands; however, the design of communication systems at these extremely high frequencies is very challenging. It will take some years for products of reasonable cost and satisfactory performance to be introduced in the market. Also, adding bandwidth does not address the problem at its root. What is needed is a broadband, interference-free, or at least interference-resistant technology, allowing easy frequency reuse made available to the customer at an affordable cost [4]. Considering the rapidly growing wireless consumer devices, it is evident that the need for such technology is quite urgent.

The wireless handheld devices require ever-increasing bandwidth, and along with that, explosive growth in interdevice wireless communications is already creating huge demands on spectrum resources, which can be resolved only by near-zero-sum allocation decisions, made through a mixture of bidding and politics.

In economy, the game theoretic Nash equilibrium (named after John Forbes Nash, who proposed it) [6] is a solution concept of a game involving two or more players, in which each player is assumed to know the equilibrium strategies of the other players, and no player has anything to gain by changing only his own strategy unilaterally. If each player has chosen a strategy and no player can benefit by changing his or her strategy while the other players keep theirs unchanged, then the current set of strategy choices and the corresponding payoffs constitute Nash equilibrium. The practical and general implication is that when players also act in the interests of the group, then they are better off than if they acted in their individual interests alone.

Unfortunately, with spectrum usage, Nash equilibrium may result in a spectrum crunch [2], if the participants do not cooperate. An example of this was the Cellular Digital Packet Data (CDPD). This was a wide-area mobile data service, which used unused bandwidth normally used by AMPS mobile phones between 800 and 900 MHz to transfer data. Speeds up to 19.2 Kbps were possible. The service was discontinued in conjunction with the retirement of the parent AMPS service; it has been functionally replaced by faster services such as 1xRTT, EV-DO, and UMTS/HSPA. Developed in the early 1990s, CDPD was large on the horizon as a future technology. However, it had difficulty competing against existing slower but less-expensive Mobitex and DataTac systems, and never quite gained widespread acceptance before newer, faster standards such as GPRS became dominant. CDPD had very limited consumer offerings. Though AT&T Wireless first offered the technology in the United States under the PocketNet brand, they eventually refused to activate the devices. Despite its limited success as a consumer offering, CDPD was adopted in a number of enterprises and government networks. It was particularly popular as a first-generation wireless data solution for telemetry devices (machine-to-machine communications) and for public safety mobile data terminals. In 2004, major carriers in the United States announced plans to shut down CDPD service. In July 2005, the AT&T Wireless and Cingular Wireless CDPD networks were shut down. Equipment for this service now has little to no residual value [7].

Another example of co-existence with already existing services over radio spectrum (a form of bandwidth sharing) is the idea of ultra wideband (UWB) [8] that proposed to use direct-sequence spread spectrum sharing bands over 7 GHz of already allocated radio spectrum. This technology did not go too far either, although a huge amount of resources was spent on demonstrating the feasibility of the technology through research and development. The developed technologies work perfectly according to the specifications, but there is no public acceptance in adopting these techniques.

Some views on bandwidth sharing, be it through cognitive radios or dynamic spectrum allocation (DSA) [9], are given here. The wireless/mobile environment is very dynamic. To capture when a piece of spectrum is free and available (known as white space) in order to reallocate it, many accurate energy sensors have to be installed to identify these available bands. Then a command has to be sent to a cloud (database) at a distance in order to make the availability of the idle bands known to users in order to reallocate these available bands. This is a very difficult and expensive proposition in a densely populated metropolitan area where bandwidth sharing is needed the most. There might be several available portions of bands idle in rural areas; however, bandwidth and channel borrowing concepts only work over short distances. In dense metropolitan areas, by the time sensing is done and a reallocation decision is reached, the spectrum availability status may be different.

There are many practical odds against bandwidth sharing through cooperation, and there are commercial risks involved with the results (as with CDPD). Some of these are (i) cost effectiveness of sensors and the number of sensors required; (ii) willingness of spectrum resource managers (FCC, NTIA, ITU) to allow a commercial enterprise to resell spectrum and to dynamically allocate spectrum resources; (iii) ability to raise sufficient capital to deploy a network of sensors and spectrum monitoring/allocation system in a dense geographical area; (iv) and ability to sign customers onto a plan to utilize dynamically allocated spectrum.

These are the common problems with dynamic allocation of spectrum. Regulations and protocols attempt to address these but are usually difficult to construct, and even harder to enforce. Therefore, we look for viable approaches.

We need "new spectrum," and we also need mechanisms to address the "tragedy of the commons" problem [10] with the allocated spectrum. The tragedy of the commons is a dilemma arising from the situation in which multiple individuals, acting independently and rationally, consulting their own self-interest, will ultimately deplete a shared limited resource, even when it is clear that it is not in anyone's long-term interest for this to happen. This dilemma was described in an influential article titled "The tragedy of the commons," written by ecologist Garrett Hardin and first published in the *Science* journal in 1968 [10]. Therefore, unless some sort of regulation is implemented, the rational strategy for individual users never produces Pareto optimality among permitted users [6]. This optimality is also borrowed from game theory arguments, except that here the users have a self-policing or self-regulation imposed on them.

Self-regulation is achieved by utilizing higher frequency carriers. Higher frequency waves above 30 GHz tend to travel only a few miles or less and generally do not penetrate solid materials very well. This offers a sustainable solution for the current spectrum crunch [2]. Actually, the July, 1997 Federal Communications Commission's Office of Engineering and Technology in USA Bulletin #70 "Millimeter Wave Propagation: Spectrum Management Implications" [11] reads thus: "The absorption bands (e.g., at 23 GHz or 60 GHz) would be applicable for high data rate systems where secure communications with low probability of intercept is desirable; for services with a potentially high density of transmitters operating in proximity; or for applications where unlicensed operations are desirable."

To address the spectrum scarcity problem in current wireless systems, we are examining the concept of adaptive rate delivery of future mobile and portable multimedia services with high bit rates (>100 Mbps) for localized areas [4]. The motivation for operators of such bands to actually choose to self-limit is that by doing so, they improve the signal-to-noise ratio against competing users at a lower cost than trying to overcome interference. These characteristics of wave propagation are not necessarily disadvantageous as they enable more densely packed communications links. Thus, high frequencies can provide very efficient spectrum utilization through "selective spectrum reuse," and naturally increase the security of transmissions. Hence, OWC is a direct solution in the spectrum reuse scenario.

Two branches of optical wireless have emerged contemporaneously. In one branch, semiconductor LED is considered to be the future primary lighting source for buildings, automobiles, and aircrafts. LED provides higher energy efficiency compared to incandescent and fluorescent light sources, and it will play a major role in the global reduction of carbon dioxide emissions, as a consequence of the significant energy savings. Lasers are also under investigation for similar applications. These core devices have the potential to revolutionize how we use light, including not only for illumination but also for communications, sensing, navigation, positioning, surveillance, and imaging. The second branch uses coded optical signals within two coherent optical side bands centered at different wavelengths. The two sidebands, at least one of which carries a message, are transported over long distances to a broadcast station, at which point, heterodyne interference of light within the two bands produces an electromagnetic wave at microwave or millimeter wave frequencies that is modulated by the lower frequency optically coded message. The electromagnetic wave carrying the coded message is then broadcast by an antenna. A wireless receiver can reply wirelessly over a return path via an electrically generated wave carrying an electrically generated coded message. Wired optical networks and various wireless networks are thus merged. Each of the optical wireless networks briefly described earlier has its unique applications, message coding, security features, and technology for sending and receiving messages. Among applications in this area are multiband, multiservice wireless over optical access, distributed radio-over-fiber access network for cloud-computing, broadband millimeter-wave wireless sensor communications, and microwave photonics for integrated multigigabit wireless systems.

Visible light and infrared light (IR) exhibit very similar qualitative behavior because of the closeness of their wavelengths; however, in terms of indoor communications, only IR has been used mostly until now. The reason is that until recently, it was not possible to manufacture highly efficient WLEDs. As LEDs increasingly displace incandescent lighting over the next few years, general applications of VLC technology are expected to include wireless Internet access, vehicle-to-vehicle communications, broadcast from LED signage, machine-to-machine communications, positioning systems, navigation, and so on. The VLC technology has potential in a number of specialized application areas including the following: (i) Indoors/Outdoors Light Positioning System (LPS) in analogy to GPS; (ii) Light Navigation Systems; (iii) Hospital and Healthcare—enabling mobility and data communications in hospitals; (iv) Hazardous Environments—enabling data communications in environments where RF might be potentially harmful (i.e., Oil and Gas, Petrochemicals and Mining); (v) Commercial Aviation—enabling wireless data communications such as in-flight entertainment and personal communications; (vi) Corporate and Organizational Security—enabling the use of Wireless Networks in applications where Wi-Fi presents a security risk; (vii) Wi-Fi Spectrum

Relief—providing additional bandwidth in environments where unlicensed communication bands are congested; (viii) Defense and Military Applications—enabling high data rate wireless communications within military vehicles and aircrafts; (ix) Underwater communications—between divers and/or remote-operated vehicles.

Examples of localized areas could be classrooms, hotel rooms, future homes, shopping malls, waiting rooms in airports and train stations, planes, space-crafts, and so on. Consider the area of home networking—when in the very near future, every home will be illuminated with bright visible LED lights, they can also be used as a broadband communications carrier. Light-waves at visible and IR wavelength range and beyond are confined to the walls in a room and generally do not penetrate solid materials. Hence, practical and usable networks can be readily realized, which utilize this self-limiting link distance. We call such systems high-bandwidth islands that employ this property. The motivation for operators to actually choose to transfer data through this optical band is that by doing so, the entire huge bandwidth can be reused next door, free of interference.

In large open environments where individual users may require 100 Mbps speed or more, optical wireless (OW) is a more sensible solution because of its limited cell size. Today's RF LANs realistically cannot support more than a couple of high capacity users per cell, which is highly wasteful. Multiple high-capacity users require multiple cells and thus create a situation where the cells almost completely overlap, which then raises concerns with regard to interference, carrier reuse, and so on. In contrast, OW could deliver the necessary capacity to each user through multiple user-sized cells, and because of the intrinsically abrupt boundary of these cells, interference would be negligible and carrier reuse would not be an issue. These cells, or high-bandwidth islands, can indeed solve much of the spectrum shortage problem by transferring the high-bandwidth multimedia payloads to wireless optical carriers from radio frequency. Also OW is a future proof solution, as additional capacity far beyond the capabilities of radio could be delivered to users as their needs increase with time.

VLC could be a viable option for optical wireless systems as LEDs can be used as a wireless communications transmitter. This is not possible for any other kind of lamps in broadband transmissions. One can use the same visible light LEDs not only for lighting homes but also as light sources for wireless in-house communications [12], and there is now an IEEE standards committee addressing the issues of this application. A full duplex operation thus can be realized by using IR as uplink and visible light LEDs as downlink. Using this new and developing technology along with power-line communications (PLC) and smart-grid can go a long way to mitigate the spectrum crunch problem as there will no longer be a need for separate lighting and communications equipment or interference creating RF restrictions.

It is commonly agreed that future generations of wireless communications systems will not be based on a single access technique but will encompass a number of different complementary access technologies. Surprisingly, currently perhaps the largest installed base of short-range wireless communications links are optical, rather than RF. Indeed, "point and shoot" links corresponding to the Infra-Red Data Association (IRDA) standards are installed in 100 million devices a year, mainly remote controls. It is argued that OW has an important part to play in the wider 5G vision as the communications technology of the future. Thus it is high time that multimedia transmissions requiring high-bandwidth in indoors be shifted to optical bands as an effective strategy to overcome the spectrum shortage problem.

1.2 Organization

In Chapter 2, we discuss some fundamentals of OWC systems. There are differences between radio frequency-based and OWC systems that we provide details on. We also discuss optical transmitters and receivers.

The goal of Chapter 3 is to establish a proper channel modeling method to ensure that modeling can be done accurately and fast. Channel modeling is very important for optical wireless links as optical signals bounce back and forth from the walls within a room and hence the receiver receives delayed or reflected versions of the same signal. As this is the cause of intersymbol interference (ISI) at high data rates, modeling the channel to better understand this multipath phenomenon is an important topic.

Our objective in Chapter 4 is to analyze various channel properties using the models obtained from channel modeling techniques. We discuss different topics related to indoor optical wireless channels such as root mean square delay spread and path loss. We see the effects of additional room furniture on these parameters also compared to an empty room.

Chapter 5 consists of several fundamental researches on multiple sources VLC. Source layout is one of the most important factors that affect overlapping of light footprints and thus produce ISI. It determines the pattern and extent of the overlapped lights. We explore VLC performance in conventional household layouts and investigate the impact of these layouts on VLC.

Orthogonal frequency division multiplexing (OFDM) is currently being used predominantly in RF mobile broadband communication systems because of its ability to combat ISI and robustness against frequency-selective fading caused by multipath wireless channel. OFDM is also being considered as a candidate for VLC as it offers robustness against multipath, caused by diffuse indoor OW channel. However, OFDM suffers from certain disadvantages such as high peak-to-average power ratio (PAPR). Also, optical wireless transmissions require the modulating signal to be unipolar. In Chapter 6, we develop some techniques to reduce high PAPR in OFDM-based OW systems as the nonlinear characteristics of LED transmitters can severely affect system performance. We look into various precoding-based PAPR reduction techniques. We then analyze performance of various OFDM-based OW schemes in multipath diffuse indoor wireless channels. We compare the performance of conventional schemes with a precoded version.

In Chapter 7, multiple-input and multiple-output (MIMO) techniques are included as they provide either reliability improvement or bandwidth efficiency increase. Based on these investigations, we further explore VLC performance in real applications, such as aircraft cabin wireless communications in Chapter 8.

In Chapter 9, we discuss multispot diffusing configuration where multiple spots on the ceiling of a room are created with the help of a holographic diffuser. We also discuss angle-diversity receivers that are set up in such a way that several photodiodes are arranged at different angles relative to each other and thus face different directions, or can be constructed using holographic mirrors.

In Chapter 10, we discuss an important application of VLC-based OWC techniques—indoor positioning and navigation systems. Indoor positioning has become an attractive research topic in the past two decades. However, no satisfying solution has been found with consideration to both accuracy and system complexity. Recently, research on visible light communications has offered new opportunities in realizing accurate indoor positioning with relatively simple system configurations. In this chapter, we also investigate several fundamental research topics of indoor positioning systems based on VLC technology.

References

[1] S. Arnon, J. R. Barry, G. K. Karagiannidis, R. Schober, and M. Uysal, Advanced Optical Wireless Communication Systems. Cambridge University Press, Cambridge, 2012.

[2] D. Goldman, Sorry, America: Your Wireless Airwaves Are Full. CNNMoneyTech, New York, 2012.

[3] M. Kavehrad, "Broadband room service by light," Scientific American, vol. 297, no. 1, pp. 82–87, 2007.

[4] M. Kavehrad, "Sustainable energy-efficient wireless applications using light," IEEE Communications Magazine, vol. 48, no. 12. pp. 66–73, 2010.

[5] Cisco, "Cisco Visual Networking Index: Global Mobile Data Traffic Forecast Update 2014–2019 White Paper," [Available Online]: http://www.cisco.com/c/en/us/solutions/collateral/service-provider/visual-networking-index-vni/white_paper_c11-520862.html (accessed April 17, 2015).

[6] M. J. Osborne and A. Rubenstein, A Course in Game Theory. MIT Press, Cambridge, 1994, p. 7.

[7] A. Salkintzis, "Radio resource management in cellular digital packet data networks," IEEE Personal Communications, vol. 6, no. 6, pp. 28–36, 1999.

[8] S. Wood and R. Aiello, Essentials of UWB, The Cambridge Wireless Essentials Series, Cambridge University Press, Cambridge, 2008.

[9] J. Zhu and K. J. Ray Liu, "Cognitive radios For dynamic spectrum access—dynamic spectrum sharing: a game theoretical overview," IEEE Communications Magazine, vol. 45, no. 5, pp. 88–94, 2007.

[10] G. Hardin, "The tragedy of the commons," Science, vol. 162, no. 3859, pp. 1243–1248, 1968.

[11] Office of Engineering and Technology Bulletin #70, Millimeter Wave Propagation: Spectrum Management Implications. Federal Communications Commission, Washington, DC, 1997.

[12] Short-Range Wireless Optical Communication Using Visible Light, IEEE Standard 802.15.7, 2011.

2

Fundamentals of Optical Wireless Communications

2.1 Introduction

Optical wireless communications (OWC) involves transmission and reception of signals where the carrier frequency lies in the optical domain. Whereas in radio frequency (RF)-based communication systems the carrier frequency can be from 30 MHz to 5 GHz, and for satellite and other pointed communication systems the carrier frequency can be up to 300 GHz, optical frequency ranges begin beyond the so-called THz regime. Because of the very high frequencies involved, usually in optical communications the carrier is denoted not by its frequency, but by its wavelength. Thus in RF communication systems the carriers have wavelengths from kilometer to millimeter range; whereas in optical communications, the carriers have wavelengths in the micrometers and nanometer range. Figure 2.1, as reported in Ref. [1], shows wavelengths and frequencies of all radio and optical carriers.

This major difference in the carriers involved in radio and optical communication systems leads to other dissimilarities in their implementations. For example, antennas used in RF systems for both transmission and reception are completely replaced in optical systems by sources that emit light and receivers that detect light. Other parts of the communication systems that are required in RF-based systems for transmission, such as power amplifiers, are replaced by circuits that drive the light sources in optical communications. These differences have consequences that make radio and optical propagation channels as well as the whole communication systems quite different from one another. Of course, there are similarities too, such as baseband modulation schemes, which are the same for both, though complex baseband modulation schemes, where the result of the baseband modulation is complex, cannot be used in optical communications.

In this chapter, we delve into these aspects of OWC and discuss these features in more detail. We discuss optical wireless channel characteristics along with source types and

Short-Range Optical Wireless: Theory and Applications, First Edition. Mohsen Kavehrad,
M. I. Sakib Chowdhury and Zhou Zhou.

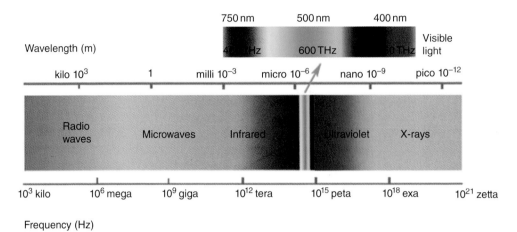

Figure 2.1 Wavelengths and frequencies of radio and optical carriers.

receiver configurations. We show that multipath fading, a property in RF communications channels, is absent for optical wireless channels. Multipath reflections, on the other hand, are present and the result is intersymbol interference (ISI) in high-speed communications. Not all source types can support all data rates. We distinguish between two main source types, namely, lasers and light-emitting-diodes (LEDs), and discuss their common features and differences. Receivers in optical communications systems, which are photodetectors, are also a very integral part in determining many performance metrics. Hence, we discuss photodetectors and their related parameters.

2.2 Communications Blocks in an OWC System

A complete block diagram of a communication system that uses optical carriers is shown in Figure 2.2. Several of the blocks perform in a similar manner as in an RF-based communication system. Bits obtained from a source are first source encoded to compress the data in order to transmit more efficiently. Examples of source encoding are speech encoding for real-time audio transmission and reception, audio encoding for offline storage that includes MP3, AAC, and other audio formats, encoding of still images such as JPEG, PNG, and other image formats and encoding of video data such as MPEG or H.264 formats. The next block is the channel encoder that adds redundant bits to the data so that at the receiver end the data can be recovered even if there is corruption due to noise and channel conditions. The channel encoder is an important part of RF-based communication systems where bit-error-rate (BER) of 10^{-6} after the demodulator block at the receiver side is acceptable only because it is assumed that channel encoders and decoders are present that will correct the errors. Channel encoding and decoding techniques as Reed–Solomon codes, low-density parity check (LDPC) codes, Turbo codes and others are used nowadays in various RF wireless standards. In an optical wireless channel these two blocks—source encoder and channel encoder—can be used without any modification from RF-based systems. One other block—the interleaver block—is not shown in Figure 2.2, but it can be added after the channel encoder block. The purpose of the interleaver is to prevent the channel decoder from failing

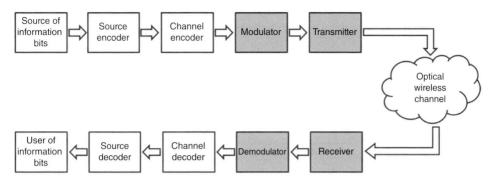

Figure 2.2   An optical wireless communication system block diagram.

due to burst errors caused by channel conditions. The corresponding block at the receiver side is the de-interleaver block that is placed before the channel decoder block. Interleavers are necessary in RF-based communication systems and are frequently used. In optical wireless channels, depending on applications, if burst errors are expected, interleavers and de-interleavers can be added.

Differences exist among RF-based and optical communication systems in the next two blocks of both the transmitter and the receiver chains, which are shaded to point out this fact. Modulators in RF-based systems can work with any modulation scheme that can produce real or complex baseband signals. However, as we will see later, optical wireless transmission process can be modeled as baseband transmission, and for the transmitter and the receiver, a process called intensity modulation/direct detection (IM/DD) is employed, which prohibits complex signals to be input to the transmitter, and at the receiver end the output is always real. This means the modulator and the demodulator blocks work with only real-valued signals. This has some advantages in terms of complexity, but it reduces achievable higher spectral efficiency that can be attained with modulation schemes that produce complex constellations such as M-QAM, M-PSK, and so on. Usually, in optical communications, higher order modulation schemes such as M-PAM or L-PPM can be employed as they produce real-valued baseband signals. Multicarrier modulation schemes such as OFDM are also used in optical communications, but again there is a difference in that the output of the OFDM modulator has to be real, which forces the input symbols to the OFDM block to be Hermitian symmetric, reducing spectral efficiency. Moreover, output of OFDM is bipolar, which has to be converted to a unipolar signal for optical transmission, which is done by either DC-biasing or clipping the negative parts of the signal. All these lead to lower spectral efficiency in optical OFDM than RF counterparts.

The transmitter block is completely different in optical wireless systems compared to RF counterparts as explained earlier. Instead of antennas that radiate electromagnetic waves in the MHz–GHz range, transmitters in optical wireless systems emit light. Figure 2.3 shows two types of light-emitting sources that are commonly used for OWC. Lasers and LEDs produce emissions of light with very different characteristics, and we discuss their structures and operation principles in later subsections. Both of these devices require a driver circuit that produces the current proportional to the modulated electrical waveform, which in turn drives lasers or LEDs and changes the intensity of light emitted from them. This is the basic principle of IM. We discuss more about IM shortly.

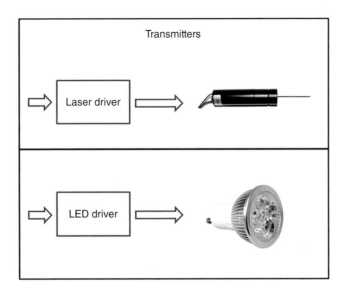

Figure 2.3   Transmitters used in optical wireless systems.

After the light radiation is emitted to an optical wireless channel, it has to be detected. Radiation detectors, that is, receivers in an OWC system are usually termed *photodetectors*. Though there are various kinds of photodetectors available, we first show here that because of the large detector area, multipath fading does not occur in an optical wireless channel. Secondly, output of a photodetector is current, which is proportional to the intensity of light that radiates on the active detection area of the photodetector. The value of this current can be higher for some photodetectors that are termed *avalanche photodiodes*, though usually cheaper photodetectors that produce a lower amount of current are also in use. We delve into more details of photodetectors in this chapter.

2.3   Intensity Modulation/Direct Detection (IM/DD)

In an IM transmission system, the intensity of the light emitting from sources is varied according to some characteristics of the modulating signal. Usually, the amplitude of the modulating signal is taken as the property according to which the instantaneous optical power output is varied. There are some consequences to this scheme that are different from RF antenna-based transmission systems. As the radiant intensity is varied according to the amplitude of the modulating signal, and intensity is a real-valued physical parameter, the modulating signal also has to be real-valued, that is, the modulating signal, if it is in baseband, has to be real. It is, of course, possible to use a complex baseband modulating signal, and as in RF-based systems, introduce a higher frequency carrier wave so that the end result is real and can be used to vary the intensity of light. However, it is generally not implemented in optical wireless transmissions because the high frequency RF carrier will require the light-emitting source to have higher bandwidth for intensity modulation. Similarly, the receiver photodetector will also require higher bandwidth for detection of this higher frequency RF carrier. Thus, when intensity modulation is specified, it is usually implied that the modulating signal is in real baseband form.

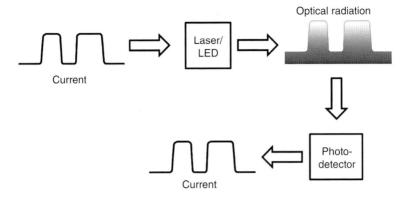

Figure 2.4   Basics of an IM/DD system.

DD is applicable to all photodetectors in optics, where a proportional current or voltage is produced to the intensity of light that radiates on the active detection area of the photodetector. It is a noncoherent form of detection where the frequency or wavelength of the light radiation is not considered in the detection. Photodetectors have different sensitivities to different wavelengths of light, and hence, some dependence on wavelength still exists, but it is unlike coherent detection where a local oscillator produces a carrier similar to the incident radiation. In optical communications, coherent detection is possible, though it is complex and applicable to very specific applications. For general indoor optical wireless applications, a photodetector that employs direct detection is used most frequently. Figure 2.4 illustrates an IM/DD system for simplicity and cost reasons.

2.4   Optical Transmitters

As we have mentioned in the previous subsections, two main branches of optical transmitters are in use for optical wireless transmissions—lasers and LEDs. We discuss in brief their main operation principles:

1. *Lasers*: The word "LASER" is an acronym which stands for light amplification by stimulated emission of radiation. The acronym describes the operational basics of how lasers are created. Lasers are unique sources that can simultaneously produce both coherent, that is in-phase radiation and monochromatic wavelengths [2]. The propagation characteristics of lasers enable them to be applied in various applications that are not possible with spontaneous emission sources, for example, fine definition information transfer as in compact disk players and laser printers, high energy concentration applications in medical surgery, metal cutting, military applications, and so on. Additionally, laser beams travel long distances with minimum dispersion if properly guided, which is a very important characteristic and because of which lasers are used in communications and metrology applications. Measuring the distance from the earth to the moon by directing a laser beam to the moon and bouncing it back was possible due to this characteristic. The monochromatic nature of laser has made it applicable for optical fibers. Because lasers can produce a single wavelength, technologies such as dense wavelength division multiplexing

(DWDM) have been made possible. Application of lasers in OWC is hence important. As lasers are emitted in a single beam from laser sources that are coherent and focused (usually by the use of a collimator lens), the energy emitted from the source becomes an important factor to consider in indoor wireless communications because high energy laser emissions are harmful to eyes. There are lasers for different wavelengths, ranging from visible to infrared. For different wavelengths, the highest energy that a laser source can emit but still not be harmful to eyes is also different. Laser safety standards, for example [3], define exactly the amount of energy per second, that is, power that a laser source in free space is permitted to have. There are classifications on the energy ratings of lasers, such as class I, class II, and so on. According to standards, class I lasers are those that do not pose a hazard to eye-safety at any condition.

2. *LEDs*: These are definitely one of the most popular opto-electronics sources. They are inexpensive compared to lasers and consume little power. The operation principles of how LEDs emit radiation can be described by first pointing out that LEDs are basically semiconductor junctions. All semiconductor diodes produce radiation when electrons from the conduction band recombine with the holes in the valence band. In a normal silicon diode, this radiated wavelength is absorbed by the surrounding material and cannot escape the junction. An LED is constructed such that the semiconductor has a high-energy gap and the junction is constructed so that the radiation from the junction can escape [4]. LEDs for different wavelengths exist, ranging from visible to infrared. In the early days of optical wireless research, demonstrations usually employed infrared LEDs. Nowadays when transmitters use infrared wavelength, usually lasers are employed instead. LEDs are mainly used in the visible wavelength range, that is for visible light communications. The main difference between a laser and an LED is that LEDs produce light with a broad spectrum, that is, LEDs are not monochromatic as lasers. Also, the emitted light is non-coherent, that is, the carriers are not in-phase. Another significant difference is that lasers have a very large bandwidth, that is, a laser source can be modulated with very high-speed modulating signals, while LEDs usually have a bandwidth of about 20–100 MHz. The output from a laser source is usually coupled to an optical fiber, which can be collimated by a lens to emit into free space, that is, the output in free space is a beam. The output of an LED source, on the other hand, has a specific radiation pattern. In the next chapter, we discuss more about radiation patterns from sources such as LEDs and radiation patterns that are created as laser beams hit a surface and get reflected.

2.5   Optical Receivers

As we discussed earlier, optical receivers are called photodetectors. The purpose of photodetectors is to produce an output current or voltage that is proportional to the intensity of incoming light on the active detection area. The most common types of photodetectors are classified as photoelectric detectors that include photodiodes and phototransistors where electrons are released as a result of photon radiation on a semiconductor surface, which can be either junction or bulk type. The junction photodiode is a *p*- and *n*-type semiconductor junction similar to the junction used in an LED. However, the function of a photodiode junction is the exact reverse to that of an LED junction. In an LED junction, photons are released as a response to the current flow through the junction; whereas in a photodiode junction, the photons are absorbed, resulting in free carriers that become current through the junction [5]. Since these types of photodetectors are inexpensive, reliable, and small, they have become the building blocks in modern optoelectronics technology. As photodiodes are receivers, it is important to understand the noise generated as light is radiated on them and absorbed. Photodiodes also have limited bandwidth, dependent upon several factors. The range of bandwidth can be from a couple of megahertz to a couple of gigahertz. The larger the bandwidth, the smaller the active detection area becomes. The reason for this is the capacitance introduced by the detection area, which has to be reduced in order to switch more rapidly. A smaller detection area signifies the necessity of receiving more optical power to produce the same amount of current that can be produced by a photodiode with a larger detection area. Hence, there is a trade-off in the sense that in order to increase the bandwidth of a photodiode, the required light intensity has to be increased to produce the same amount of current.

We now define some important parameters regarding photodiodes, and show by an example of a photodiode with specifications from its manufacturer, the amount of optical power needed for it to operate.

The first important characteristic of a photodiode is its quantum efficiency, η. It is the number of electron–hole pairs that are generated per incident photon of energy $h\nu$ and can be given by

$$\eta = \frac{\text{No. of generated electron} - \text{hole pairs}}{\text{No. of photons incident on photodiode}} = \frac{I_p / q}{P_0 / h\nu} \tag{2.1}$$

where I_p is the average current output from the photodiode and q is the charge of electron. Hence, I_p/q gives the number of electron–hole pairs n_e that are generated in the junction as $I_p = n_e q / t$ for some time t. Each photon carries an energy of $h\nu$, hence the average power P_0 carried by n_p photons during time interval T is $P_0 = n_p h\nu / T$. In a practical photodiode, 100 photons will create between 50 and 95 electron–hole pairs. This gives a quantum efficiency ranging from 50 to 95%.

The photodiode responsivity $\mathcal{R}$, which is one of the parameters by which a photodiode performance can be characterized, has a simple relation with quantum efficiency, given by

$$\mathcal{R} = \frac{I_p}{P_0} = \frac{\eta q}{h\nu} \tag{2.2}$$

Photodiodes whose junction area is comprised of InGaAs can have a responsivity as high as 1 A/W or even very slightly higher than that at a wavelength of 1550 nm. Responsivity varies with the wavelength and material used. Figure 2.5 shows a sample responsivity curve from a photodiode (DET08CFC) manufactured by Thorlabs Inc. As it can be seen, at 1550 nm, the responsivity is the highest, 1.05 A/W and it degrades fast as the wavelength becomes smaller. Hence, it is important to select a proper photodiode based on the wavelength chosen for the communications system that has the highest responsivity at that wavelength. This is true for monochromatic systems. If an LED is used as the source, the output current of the photodiode will be dependent on a broad range of wavelengths; hence, it is best to select a photodiode whose responsivity curve is flat over the wavelengths generated by the LED.

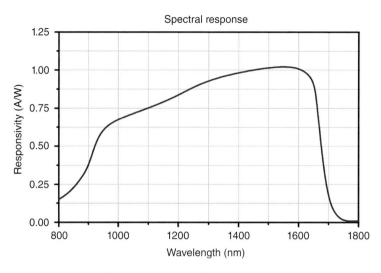

Figure 2.5 Responsivity curve of a photodiode as an example.

For measuring the bandwidth of a photodiode, the rise time and the fall time are important parameters. The bandwidth of a photodiode puts a limit on the achievable data rate of a communications system; hence, knowledge of the bandwidth of a photodiode that is to be used in the receiver is important. Rise time is the measure of the time response of a photodiode to a stepped light input, and is defined as the time required for the output to change from 10 to 90% of steady output level. Fall time is a similar measure, except it is the time required for the output to change from 90 to 10% of steady output level at the falling edge of stepped light input. The bandwidth of a photodiode is related to fall time and rise time. The rise time and the fall time of a photodiode indicate that there is a gradual drop in the output level beyond a certain frequency. The point at which the output has dropped to 50% of its low-frequency value is called the 3-dB point. At this point only half as much optical power is converted to current compared to lower frequencies. The 3-dB point defines the receiver bandwidth. If the rise and fall times are equal, the 3-dB bandwidth can be estimated from the rise time t_r by

$$f_{\text{BW}} = \frac{0.35}{t_r} \tag{2.3}$$

There are two types of noise currents at the output of a photodiode—dark current noise and quantum shot noise. The relatively small current through the photodiode in the absence of light is called dark current. It is also referred to as reverse bias leakage current in nonoptical devices and is present in all diodes. Dark current is generated due to the random generation of electrons and holes within the depletion region of the device. Dark current must be accounted for by calibration if a photodiode is used to make an accurate optical power measurement. It is also a source of noise when a photodiode is used in an optical communication system—either wireless or wired (optical fibers)—as dark current is always present, even when no photons enter the active detection area. Shot noise arises from the statistical nature of the production and collection of photoelectrons. To describe the origin of shot noise a bit more in detail, we can denote the classical light power P_0 as the result from an average over a few optical cycles. Fluctuations in P_0 are transferred to the photocurrent I_{ph}. The ideal classical optical signal is assumed to exhibit a constant amplitude and phase, and thus no fluctuation is expected in the photocurrent. However, according to quantum mechanics, ideal optical signal consists of a sequence of independent photons that are Poisson distributed in time. Each photon generates an electron–hole pair with probability η, which is actually the quantum efficiency described earlier. So, the photocurrent I_{ph} consists of a stream of statistically independent elementary charges which are Poisson distributed in time. This type of noise is called the shot noise. Of course, a photodiode must be connected to a resistor which is also the source of thermal noise. The noise currents are given by

$$\text{Shot noise, } i_{shot} = \sqrt{2qI_{ph}B} \tag{2.4}$$

$$\text{Noise due to dark current, } i_{dark} = \sqrt{2qI_dB} \tag{2.5}$$

$$\text{Thermal noise current, } i_{thermal} = \sqrt{\frac{4kTB}{R_{load}}} \tag{2.6}$$

where B is the bandwidth of the receiver operation, I_d is the average dark current, k is the Boltzmann's constant $(1.38 \times 10^{-23} \, \text{JK}^{-1})$, T is the absolute temperature, and R_{load} is the resistor connected to the photodiode. The other terms are the same as explained in previous equations. The total noise current can be given by

$$i_{noise}^2 = i_{shot}^2 + i_{dark}^2 + i_{thermal}^2$$
$$\Rightarrow i_{noise} = \sqrt{i_{shot}^2 + i_{dark}^2 + i_{thermal}^2} \tag{2.7}$$

If we consider only the noise for which the photodiode is directly responsible, we may omit the thermal noise from (2.7), and the noise current becomes

$$i_{noise} = \sqrt{2q(I_d + I_{ph})B} \tag{2.8}$$

One important parameter that is directly related to the noise currents just described is the noise equivalent power (NEP). NEP is defined as the optical signal power required to generate a photocurrent I_{ph} that is equal to the total noise current i_{noise} at the photodiode at a given wavelength and within a bandwidth of 1 Hz, that is, NEP represents the required optical power to achieve an SNR of 1. NEP is essentially the minimum detectable power. As noise levels are proportional to the square root of bandwidth, NEP is also specified as power per square root of bandwidth $(\text{WHz}^{-1/2})$. We can derive some simple equations based on this definition such as

$$\text{SNR} = \frac{I_{ph}^2}{i_{noise}^2} = 1 \tag{2.9}$$
$$\Rightarrow I_{ph} = i_{noise} = \sqrt{2q(I_d + I_{ph})B}$$

Thus,

$$\text{NEP} = \frac{P_{\text{NEP}}}{\sqrt{B}} = \frac{I_{\text{ph}}}{\mathcal{R}\sqrt{B}}$$

$$= \frac{1}{\mathcal{R}}\sqrt{2q\left(I_{\text{d}}+I_{\text{ph}}\right)} \tag{2.10}$$

We can further evaluate the photocurrent by

$$I_{\text{ph}} = \sqrt{2q\left(I_{\text{d}}+I_{\text{ph}}\right)B}$$

$$\Rightarrow I_{\text{ph}} = Bq + \sqrt{B^2 q^2 + 2qI_{\text{d}}B} \tag{2.11}$$

where we have discarded the solution that yields negative results with practical values. Hence, NEP becomes

$$\text{NEP} = \frac{1}{\mathcal{R}}\sqrt{2q\left(I_{\text{d}}+I_{\text{ph}}\right)}$$

$$\Rightarrow \frac{P_{\text{NEP}}}{\sqrt{B}} = \frac{1}{\mathcal{R}}\sqrt{2q\left(I_{\text{d}} + Bq + \sqrt{B^2 q^2 + 2qI_{\text{d}}B}\right)} \tag{2.12}$$

$$\Rightarrow P_{\text{NEP}} = \frac{\sqrt{B}}{\mathcal{R}}\sqrt{2q\left(I_{\text{d}} + Bq + \sqrt{B^2 q^2 + 2qI_{\text{d}}B}\right)}$$

Let us plot this equation with real parameters from the specification sheet of the photodiode DET08CFC manufactured by Thorlabs Inc. The mentioned photodiode has a rise time and fall time of 70 ps, and thus a bandwidth of about 5 GHz. The peak responsivity at 1550 nm is 0.9 A/W, and the dark current is 1.5 nA. Charge of electron q is 1.6×10^{-19} C. Hence, by placing these values into (2.12) we can obtain the minimum required optical power for detection at different bandwidth values. The plot is shown in Figure 2.6. It can be seen from the plot that the minimum optical power required for detection at the highest operating bandwidth of 5 GHz for this photodiode is about 19.5 nW. If we had started with a photodiode having a larger bandwidth, greater than 5 GHz, we would find the required minimum optical power to be larger still. Hence, it is important to ensure that the minimum amount of optical radiation reaches the photodiode active detection area in an optical communication system for the receiver to work.

2.6 Optical Wireless Channel Propagation Characteristics

Now that we have described the basics of optical transmitters and receivers, it is possible to describe optical wireless channel characteristics in a more detailed fashion. Let us first describe the phenomenon which we have mentioned earlier: that there is no multipath fading in an optical wireless channel. There are always some multipath effects present in optical wireless channels due to reflections of light from any reflecting surface that reaches the receiver. However, unlike RF wireless channels, this does not lead to multipath fading. In RF wireless channels, due to multipath effects, cancellations, and additions of multipath

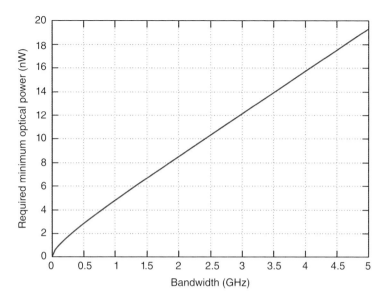

Figure 2.6   Required minimum optical power for detection vs. receiver operating bandwidth for an example photodiode.

components occur as they reach the receiver antenna which is the source of multipath fading. Let us assume there is no temporal variation in the channel and also let us ignore noise for the time being, as we want to show how multipath fading does not occur in optical wireless channels. The signal transmitted through the channel is represented as

$$s(t) = \text{Re}\left\{ x(t)e^{j\omega_0 t} \right\}$$ (2.13)

where $x(t)$ is the baseband signal and ω_0 is the carrier frequency in radians. The received signal can be represented as

$$y(t) = \text{Re}\left\{ \rho(t)e^{j\omega_0 t} \right\}$$ (2.14)

where $\rho(t)$ is the sum of all multipath components, and is given by

$$\rho(t) = \sum_{k=0}^{N-1} a_k x(t - t_k)e^{j\theta_k}$$ (2.15)

where a_k, t_k, and θ_k are the amplitude, time delay, and phase of the k-th multipath component, respectively, and there are N multipath components. If we transmit a constant envelope signal, we can set $x(t) = 1$, then the received signal at some point in space becomes

$$y(t) = \text{Re}\left\{ \sum_{k=0}^{N-1} a_k e^{j(\omega_0 t + \theta_k)} \right\}$$ (2.16)

The resultant envelope and phase of the received signal can be given by

$$a = \left| \sum_{k=0}^{N-1} a_k e^{j\theta_k} \right| \tag{2.17}$$

$$\Theta = \arg \sum_{k=0}^{N-1} a_k e^{j\theta_k} \tag{2.18}$$

In (2.16), the phase of the received signal θ_k varies, depending on the various path lengths of the multipath components. Phase changes by 2π for each change of one wavelength. For this reason, the received signal varies spatially with significant feature variations separated by multiples of the wavelength. At frequencies of RF-based wireless communication systems, 800 MHz–2 GHz, the fluctuations of the signal space are separated by distances of about 37.5–15 cm. It is thus very much possible that antennas used in such systems are situated in a region of weak signal strength frequently. In contrast, at optical signal frequencies beyond the terahertz region, the weak signal strength regions must be closely spaced in the nanometer range. The photodiodes used in the receivers of optical wireless systems have detection areas that are much larger than this distance and span many such fluctuations. That is why these rapid variations are averaged and are not noticeable at the photodiode output. Figure 2.7 shows this phenomenon where the wavelengths are much smaller than photodiode detection area.

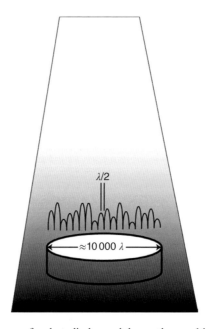

Figure 2.7   Detection area of a photodiode much larger than multipath fading fluctuations.

This is actually a major difference between optical wireless and RF-based communication systems regarding the size of the antenna. In RF wireless systems, the size of the antenna is small in comparison with the wavelength. On the other hand, a photodiode having a detection area of $1\,cm^2$ is much larger than the optical wavelengths. The result of this is that the total received power will remain the same when the detector is moved a couple of thousand wavelengths apart. This is in sharp contrast with RF-based wireless systems where channel conditions can change a lot if the receiver antenna is moved even by a fraction of a wavelength. The reflectors are also much larger for optical wavelengths compared to radio wavelengths.

Another difference between optical wireless receivers and RF receiver antennas is sensitivity to rotation. As RF-based transmissions are omnidirectional, if the antenna is rotated, small effects may be observed. However, as optical transmissions are mostly directional, rotating the receiver will result in a large decrease in received optical power at the detector.

We will now show that in an IM/DD-based optical wireless system, the propagation medium can be replaced by an equivalent baseband channel. Let us denote the normalized message signal as $x(t)$, that is, $-1 < x(t) \le 1$. We convert it to a positive signal that modulates the intensity of the light source. The positive signal can be given by $A[1 + \mu x(t)]$ where A is a DC-bias and μ is a constant, $0 < \mu \le 1$. We can obtain the intensity of the emitted optical signal by

$$I_T(t) = |f_T(t)|^2 = A[1 + \mu x(t)] \tag{2.19}$$

where $f_T(t)$ is the complex electromagnetic field of optical radiation. We now denote the complex field of the signal received through the k-th reflection at a point on the photodiode surface as $f_k(t)$, $k = 0, 1, \ldots, N-1$, and obtain the following

$$f_k(t) = \sqrt{\alpha_k}\sqrt{A[1 + \mu x(t - t_k)]}e^{-j\omega_0(t - t_k)} \tag{2.20}$$

where α_k is the attenuation factor that considers the inverse square distance-dependent power path loss, reflection losses and so on, ω_0 is the optical carrier frequency in radians and t_k is the time delay of the k-th multipath component. The total received signal is given by

$$f_R(t) = \sum_{k=0}^{N-1} f_k(t) \tag{2.21}$$

We can find the intensity of the received signal in the following way:

$$I_R(t) = |f_R(t)|^2 = f_R(t)f_R^*(t)$$
$$= \sum_{k=0}^{N-1}\sum_{l=0}^{N-1}\beta_{kl}(t)e^{j\omega_0(t_k - t_l)} \tag{2.22}$$

where

$$\beta_{kl}(t) = \sqrt{\alpha_k\alpha_l}\sqrt{A[1 + \mu x(t - t_k)]}\sqrt{A[1 + \mu x(t - t_l)]} \tag{2.23}$$

We can further manipulate (2.22) by dividing the N^2 terms into N terms where $k=l$ and $N(N-1)$ terms where $k \neq l$, and obtain the following:

$$I_R(t) = \sum_{k=0}^{N-1} \alpha_k A \left[1 + \mu x(t - t_k) \right] + \sum_{k=0}^{N-1} \sum_{l=0}^{N-1} \beta_{kl}(t) e^{j\theta_{kl}} \qquad (2.24)$$

$$\text{\small }_{k \neq l \;\; k \neq l}$$

where $\theta_{kl} = \omega_0(t_k - t_l)$ is very sensitive to path length changes. When path length changes by a wavelength, it changes by 2π. Now the excess path lengths $c(t_k - t_l)$ are in the order of centimeters or meters, that is tens of thousands of an optical wavelength, where c is the speed of light. Hence, θ_{kl} can be modeled as a random variable having uniform distribution in $[0, 2\pi]$ for any k and l ($k \neq l$). Thus, we have

$$I(t) = E \left[I_R(t) \right]$$

$$= \sum_{k=0}^{N-1} \alpha_k A \left[1 + \mu x(t - t_k) \right] + \sum_{k=0}^{N-1} \sum_{l=0}^{N-1} \beta_{kl}(t) \left[e^{j\theta_{kl}} \right] \qquad (2.25)$$

$$\text{\small }_{k \neq l \;\; k \neq l}$$

The expectation operation in (2.25) is equivalent to spatial integration over the photodiode's detection surface area. As the photodiode's surface spans many thousands of wavelengths, the second term in (2.24) containing the rapidly fluctuating θ_{kl} vanishes in the integration while the first term remains approximately constant. Also, the second term in (2.25) is equal to zero because of the uniform distribution in $[0, 2\pi)$. Now, we can convert the intensity variation to an electric current signal and then remove the DC-bias. We finally obtain the electric current as

$$y(t) = \sum_{k=0}^{N-1} a_k x(t - t_k) \qquad (2.26)$$

where a_k is a constant. Thus, the entire process can be modeled as a baseband transmission.

2.7 Conclusions

In this chapter, we have briefly described some basics of an OWC system. Since there are some fundamental differences between optical wireless systems and RF-based communications systems, in order to better understand them, we have first described the transmission and reception method of optical wireless systems, namely IM/DD where at the transmitter end, the intensity of the emitted light is varied according to the characteristics of the modulating signal; most commonly, the amplitude of the modulating signal is taken as the parameter to modulate the intensity of light. At the receiver side, DD implies averaging the received optical power and converting it to electrical current or voltage signal. A brief discussion about lasers and LEDs as optical transmitters and discussions on their differences are included next. Photodiodes act as receivers in optical systems, and we have delved into some detail on their characteristics, especially showing by use of an example, how much minimum optical power is required for a photodiode to be able to detect properly. Finally, through some mathematical analyses on the propagation characteristics of optical wireless channels, elaborating the difference among RF-based wireless channels and optical wireless channels, it has been shown that multipath fading is absent in the latter. Optical wireless transmission, propagation, and reception can be completely modeled as a baseband system and some basic mathematical analyses have been presented to verify it. With the basics of how optical wireless channels operate, we delve into more details of channel modeling methods in Chapter 3.

References

[1] NIST, "Optical Frequency Combs," [Available Online]: http://www.nist.gov/public_affairs/releases/frequency_combs.cfm (accessed April 17, 2015).

[2] A. E. Siegman, *Lasers*. University Science Books, Mill Valley, CA, 1986, p. 1283.

[3] *Safety of Laser Products—Part 1: Equipment Classification, Requirements and User's Guide*. International Electrotechnical Commission (IEC) Standard 60825-1, Geneva, 2007.

[4] E. Uiga, *Optoelectronics*. Prentice Hall, Englewood Cliffs, NJ, 1995.

[5] G. Keiser, *Optical Communications Essentials*. McGraw-Hill, New York, 2003.

3

Indoor Optical Wireless Channel Modeling Methods

3.1 Introduction

In Chapter 2, we have discussed some basics of transmitters and receivers of optical wireless communication systems, described intensity-modulation direct-detection (IM/DD), that is, the modulation and demodulation scheme used in an indoor optical wireless channel (OWC), and included analyses that showed indoor OWC does not have multipath fading and the end-to-end process can be modeled as a baseband transmission system. In this chapter, we will first show different configurations of indoor OWCs in terms of how sources and receivers are placed and used, and the results at the receiver ends due to this difference. We will see that channel impulse response plays a very significant role in the performance of indoor optical wireless systems, especially when data rates become high. Therefore, methods of calculating channel impulse responses become an important topic, as we will discuss and describe the difficulties involved in simulating an impulse response of an indoor OWC. Different algorithms exist that can approximately calculate the impulse response of an indoor optical wireless environment. The main focus of this chapter is on the features and performances of some of these algorithms and their comparisons.

3.2 Source and Receiver Configurations

Indoor OWC are broadly categorized as line-of-sight (LOS) and non-line-of-sight (NLOS) [1], based on whether there is a direct unobstructed optical path between the transmitter and the receiver. Both channel types can be utilized for communications, depending on different requirements of transmitter and receiver configurations. Transmitters used for downlink data streams are usually placed at a fixed location, preferably attached to the ceiling of a room or on some fixture that has been placed higher than most furniture and equipment in the room.

Short-Range Optical Wireless: Theory and Applications, First Edition. Mohsen Kavehrad,
M. I. Sakib Chowdhury and Zhou Zhou.
© 2016 John Wiley & Sons, Ltd. Published 2016 by John Wiley & Sons, Ltd.

If this is the case, the receiver can be placed anywhere in the room, and an unobstructed optical path will then exist between the transmitter and the receiver for most of the time, and the channel is denoted as LOS channel. However, due to various room configurations, the transmitter may not be placed at a high level, or even if it is, the receiver can be mobile, depending on requirements, and an unobstructed optical path may not exist then. The optical channel in this case is denoted as NLOS channel, where the photons from the transmitter are reflected off the walls of the room before reaching the receiver.

Further classification is possible in terms of whether the transmitter directs its emission toward the receiver or toward a small area of some reflective surface, for example, the ceiling, to maximize the amount of light that reaches the receiver. These cases are termed as directed links, whereas if the transmitter does not actively maintain its direction of emission toward the receiver or if the transmitter emits toward a broad surface area instead of a point, the links are said to be non-directed. Links can be further classified depending on the type of receiver employed. The receiver may have a very small field-of-view (FOV) so that it receives light only from a directed emission toward it, or the receiver may have a large FOV so that it can receive light from a broader emission.

Figure 3.1 shows possible configurations of LOS links. Figure 3.1a and b show directed LOS links where the transmitter is usually a laser beam or a very pointed LED emitter. There are differences in receiver architecture too: as shown in Figure 3.1a, the receiver has a small FOV, and as in Figure 3.1b, the receiver has a large FOV. Figure 3.1c and d illustrates non-directed LOS links where the transmitter is a diffused source, usually an LED, or a laser beam passed through a diffuser. Similar to the directed LOS links, Figure 3.1c shows a receiver with a small FOV and Figure 3.1d a receiver with a large FOV.

Figure 3.2 shows possible configurations of NLOS links. Similar to Figure 3.1, Figure 3.2a and b show directed NLOS links, but in this case, the transmitter emits toward a small area of the ceiling so that the receiver can receive light through reflections. The specific point or small area of the ceiling toward which the transmitter directs its emission is usually determined beforehand so that maximum amount of light may reach the receiver compared to other locations on the ceiling. These links are also termed as quasi-diffuse links. Figure 3.2c and d illustrate non-directed NLOS links where the transmitter emits toward the ceiling and has a broad emission pattern and the receiver receives light through reflections. These links are termed as diffuse links. Figure 3.2a and c show receivers that have a small FOV and Figure 3.2b and d show receivers having a large FOV.

For NLOS links, photons emitted from the transmitter do not directly reach the receiver; rather they travel initially in other directions determined by the emission pattern of the transmitter. As the photons hit any reflecting surface, they can be absorbed or reflected with lower energy, which is determined by the reflecting coefficient of the surface. After experiencing multiple bounces off the surfaces in the indoor OWC environment, photons reach the aperture of the receiver lens. Figure 3.3 demonstrates this process by showing a directed NLOS link where the transmitter, in this case a laser, emits a beam toward a wall where the beam is diffused. The diffused radiation from the reflection further propagates through the room and after some more reflections reaches the receiver. The figure shows a single ray, as an example, from the first reflection that undergoes two more reflections before arriving at the receiver. Of course, the receiver will also receive light from all these reflections, and thus not all rays that arrive at the receiver undergo three reflections; some rays may reach the receiver after one or two reflections. Also, in the figure, only one ray has been shown as an example, whereas in reality innumerable rays are generated from each reflection.

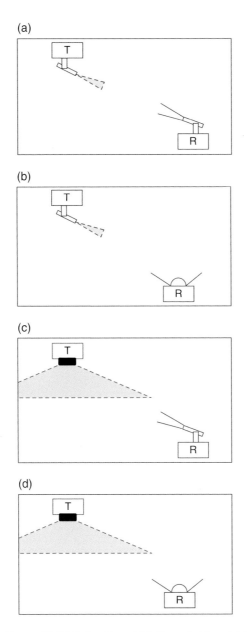

Figure 3.1 Types of LOS links: (a) and (b) directed; (c) and (d) non-directed.

Hence, in case of NLOS links, the transmitted signal experiences multipath phenomena where a pulse is broadened as photons appear with various delays. Multipath reflections can be important in some LOS channels too, where the transmitter is not a pointed laser beam, but a wide-beam LED source, and the receiver having a finite FOV will capture photons that are reflected from the walls. The effect of multipath reflection is inter-symbol-interference (ISI) because of pulse broadening and is a limiting factor on the bandwidth of the channel.

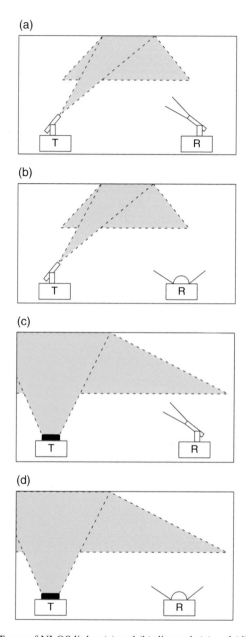

Figure 3.2   Types of NLOS links: (a) and (b) directed; (c) and (d) non-directed.

Therefore, the importance of theoretical and experimental analyses of multipath phenomena in indoor OWCs is evident. Experimentally, the frequency response of the channel is determined, and by applying inverse Fourier transform on the frequency response, the impulse response is obtained [2–4]. The bandwidth of the channel can be found from the frequency response, while some other important parameters such as average delay and

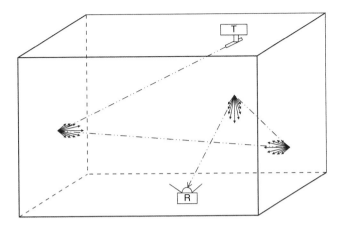

Figure 3.3   Multiple bounces of a ray before arriving at the receiver.

root-mean-square (RMS) delay spread can be calculated from the impulse response. Theoretical analyses consist of simulations where the channel is modeled by approximating the impulse response by different algorithms. This chapter will focus primarily on the algorithmic side of impulse response approximation by simulations. It should be noted here that the impulse response, and so the frequency response, of an indoor OWC, is fixed for a specific room configuration and specific positions of transmitters and receivers; that is, when a particular indoor OWC is specified, it should be understood that the channel consists of a particular room configuration along with fixed positions of transmitters and receivers. If any positions of transmitters or receivers are changed, or a slightly different room configuration is assumed, the channel will no longer be the same, and a different impulse response and frequency response will be obtained. Also, for a specific indoor OWC as just described, the system is linear time-invariant and the impulse response will completely characterize that channel.

3.3   Steps for Modeling of Indoor OWC Environment

For modeling indoor OWCs, the first step is to define a coordinate system for the environment. The coordinate system will be helpful for later steps, mainly defining positions of sources and receivers. Models that generate rays and propagate rays will also use this coordinate system extensively to describe the directions of the rays. After defining the coordinate system, sources and reflections have to be defined or modeled. Point sources that are diffused and reflections from any reflecting surface can be modeled as a generalized Lambertian emission pattern [5]. The next step is to calculate received power from LOS links that may exist between the source and the receiver. Models that divide all reflecting surfaces of the indoor environment into small reflecting elements are also heavily dependent on this step, as the small reflecting elements can be considered as receivers that receive light from the source. The NLOS portion of the impulse response has to be calculated next, which is where different models differ in their methods. In this chapter, we will describe the major algorithms and mention their variations that are found in the literature.

3.4   Models of the Room and Other Reflecting Surfaces

An indoor optical wireless environment must consist of a room, along with furniture inside the room, if necessary. Generally, a room is defined as having six reflecting surfaces, each of them perpendicular to four others. Other furniture and reflecting surfaces within the room may be modeled as boxes that have surfaces all perpendicular to each other. This property of perpendicularity is not necessary in the models and slanted reflecting surfaces can also be incorporated in the calculations. For now, to keep the description of the algorithms short and to the point, we assume there are no other reflecting surfaces than the walls, the ceiling and the floor within the room, and they are perpendicular to each other.

The convention of coordinate system is important as described in Section 3.3. Hence, to clarify the notion, Figure 3.4 shows exactly where the origin of the coordinate system is located. The figure also shows the convention followed for defining the length, width, and height of the room. The wall at the far end of the room in the positive x-direction is termed as the north wall. Hence, the wall opposite to it is the south wall. The rightmost wall is the east wall while the leftmost wall is the west wall. The ceiling and the floor are self-explanatory. Thus, all points on both the north and the south walls are on the yz plane, in addition to having $x=0$ for the south wall and $x=$room length for the north wall. Similarly, all points on both the east and the west walls are on the xz plane, in addition to having $y=0$ for the east wall and $y=$room width for the west wall. Points on the ceiling and the floor are defined similarly, that is, all points on both the ceiling and the floor are on the xy plane, in addition to having $z=0$ for the floor and $z=$room height for the ceiling.

3.5   Radiation Patterns

The source or sources that are used in different modeling algorithms are considered to be point sources. Point sources are formed when the beams of laser emitters hit a surface. Since the diameter of the beam is small, the impact area where that beam meets a surface can be considered as a point source. Generally, LED sources are not point sources as they have a finite area; however, for smaller LED sources, for example, LEDs that are used at homes to replace incandescent light bulbs, this finite area is ignored, and they are considered as point sources. Hence, there could be some errors when considering impulse response calculations simulating LED lights as point sources.

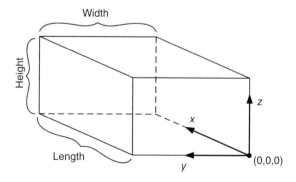

Figure 3.4   Conventions of coordinate system for a room.

3.5.1   Radiation Patterns of Point Sources

The angular distribution of radiant output power of a point source is best modeled by a generalized Lambertian pattern [5] having uniaxial symmetry:

$$dP_{S}(\Omega) = \frac{n+1}{2\pi} P_{S} \cos^{n}(\theta) d\Omega$$

$$\Rightarrow dP_{S}(\theta,\varphi) = \frac{n+1}{2\pi} P_{S} \cos^{n}(\theta) d\varphi d\theta \tag{3.1}$$

where P_S is the total optical power emitted from the source, dP_S is the optical power emitted into the solid angle $d\Omega$, the Lambertian mode number n defines the directivity of the radiation pattern, where higher values of n indicate more pointed radiation pattern and $n > 0$, θ is the angle between the normal of the source and the direction of the emitted radiation and $\theta \in [0, \pi/2]$, and φ is the angle between the plane formed by the normal of the source and the direction of the emitted radiation and the plane formed by the normal of the source and a reference axis on the surface plane of the source, $\varphi \in [0, 2\pi]$. The term $(n+1)/2\pi$ appears in (3.1) so that the total power of the source when integrated through all possible values of θ and φ equates to P_S, that is, $P_S = \int_{\theta=0}^{\pi/2} \int_{\varphi=0}^{2\pi} dP_S(\theta,\varphi)$. Figure 3.5 shows a generalized Lambertian pattern of a point source having Lambertian mode number $n = 3$.

Generalized Lambertian radiation pattern becomes more pointed as n increases. Usually for point sources, n is taken to be equal to 1. For other sources that are more pointed, n can be higher as necessary.

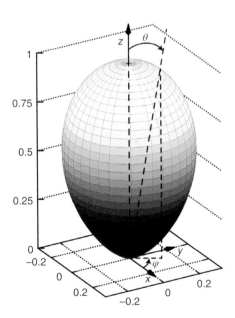

Figure 3.5   Generalized Lambertian pattern of a point source having Lambertian mode number $n = 3$.

3.5.2 Radiation Patterns of Reflections

When a ray hits a reflecting surface, a point source is generated at the impact point of the ray and the surface. This source is used to model the reflection of the incoming ray from the surface. Reflections can contain both specular and diffuse components. The characteristics of reflections from any surface depend on factors such as the material of the surface, the wavelength of the incoming ray, and the angle of incidence of the ray. The roughness of the surface relative to the wavelength is an important factor that determines the shape of the reflected radiation pattern. A surface which is smooth can act as a mirror, that is, it can reflect the incident ray in a single direction, and this direction can be calculated simply by the incident angle. A rough surface relative to the wavelength of the incoming ray reflects the ray in several directions, or in other words, there exists a radiation pattern for a rough surface. The Rayleigh criterion can be used to determine the roughness of a surface relative to the wavelength of incident radiation [6]. According to this criterion, a surface can be considered smooth if,

$$\zeta < \frac{\lambda}{8 \sin \theta} \tag{3.2}$$

where ζ is the maximum height of surface irregularities, λ is the wavelength of the incident radiation, and θ is the angle of incidence. Assuming the incident ray is normal to the surface, that is, $\theta = 90°$, and the radiation is infrared at wavelength $\lambda = 800$ nm, we have a rough surface if $\zeta > 0.1$ μm. The conclusion from this analysis is that indoor surfaces that have irregularities with height greater than 0.1 μm are rough for infrared radiation and hence the reflection patterns from these surfaces should have some diffuse components. Based on these, two models are broadly applicable to estimate the reflection pattern from indoor surfaces: (i) Lambertian reflection pattern and (ii) Phong's model [7].

3.5.2.1 Lambertian Reflection Pattern

The same Lambertian pattern that we discussed in Section 3.5.1 is applicable to model diffuse reflections. Equation 3.1 shows the generalized Lambertian radiation pattern, where for diffuse reflections, n is equal to 1. Figure 3.6 illustrates a Lambertian pattern with $n = 1$. Usually, when a Lambertian pattern is discussed without specifying any other parameters, it is understood to be a generalized Lambertian pattern with Lambertian mode number $n = 1$.

When $n = 1$, the observed radiance is the same from all directions for a surface. This can be shown as follows. In Figure 3.7, the radiance emitted from the source, in our case, a diffuse reflection, can be seen at the normal to the plane of the surface and at an angle θ to the normal. If the radiance at the normal is I Wsr^{-1} m^{-2}, the radiance at an angle θ will be $I\cos(\theta)$ Wsr^{-1} m^{-2}. Hence, the optical flux, or optical power emitted from the surface of area dA at a solid angle $d\Omega$ is $IdAd\Omega$ W at the normal of the surface and $I\cos(\theta)dAd\Omega$ W at an angle θ to the normal of the surface.

Figure 3.8 illustrates the radiance received by an observer at the normal to the surface and at an angle θ to the normal. The observer at the normal of the surface sees through an aperture of area dA_0 and the surface of area dA subtends a solid angle $d\Omega_0$ toward the observer.

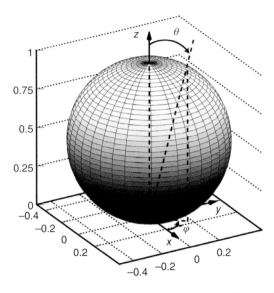

Figure 3.6 Lambertian pattern with $n = 1$ modeling a diffuse reflection.

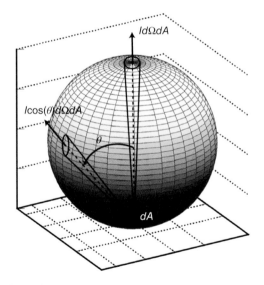

Figure 3.7 Emitted radiance from a diffuse reflection.

We assume that the aperture of area dA_0 subtends a solid angle $d\Omega$ when viewed from the surface element of area dA. Hence, the radiance observed by the user at the normal of the surface area is given by

$$I_0 = \frac{IdAd\Omega}{d\Omega_0 dA_0} \, \text{W}\,\text{sr}^{-1}\,\text{m}^{-2} \tag{3.3}$$

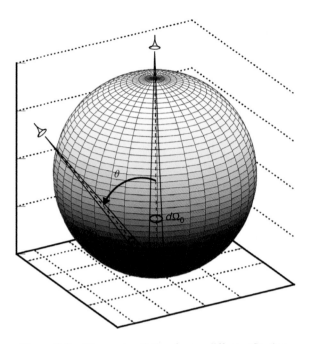

Figure 3.8   Observed radiation from a diffuse reflection.

The emitted flux from the surface element of area dA at an angle θ to its normal to a solid angle $d\Omega$ subtended by the aperture of the observer is $I\cos(\theta)dAd\Omega$ W and the solid angle subtended by the surface element toward the aperture of this observer is $d\Omega_0\cos(\theta)$. Hence, the radiance observed by the user at an angle θ to the normal of the surface area is given by

$$I_0 = \frac{I\cos(\theta)dAd\Omega}{d\Omega_0\cos(\theta)dA_0} = \frac{IdAd\Omega}{d\Omega_0 dA_0} \text{ W}\text{sr}^{-1}\text{m}^{-2} \tag{3.4}$$

Thus, the radiance observed from a diffuse reflection is the same in all directions, and hence the brightness of a surface that reflects diffusely appears to be the same when viewed from different directions. Reflections from some surfaces contain specular components as well, for which we will discuss Phong's model next. However, as measured in Ref. [5], reflections from surfaces made from commonly used materials such as paints, wood panels, textiles, and plasters are mostly diffusive, while the specular components become significant for very large angles of incidence of incoming rays.

3.5.2.2   Phong's Model

Surfaces such as varnished wood radiate a strong specular component in their reflection pattern which has been verified experimentally in Ref. [8]. The direction of the specular component follows the same law as in a mirror, that is, it is dependent on the incidence

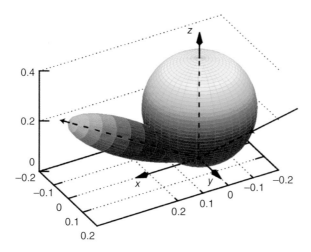

Figure 3.9 Example of a radiation pattern generated using Phong's model.

angle of the incoming ray. There is also a diffuse component in the radiation pattern as well. Such a radiation pattern can be modeled very well by Phong's model [7], which can be described by

$$R(\theta, \theta_i) = \rho P_i \left[\frac{r_d}{\pi} \cos(\theta) + (1 - r_d) \frac{n+1}{2\pi} \cos^n (\theta - \theta_i) \right] \tag{3.5}$$

where ρ is the reflection coefficient of the surface, P_i is the power of the incident ray, θ is the observation angle, that is, the angle between the normal to the surface and the observer, r_d is the percentage amount of the radiation that is reflected diffusely, n is a parameter that indicates how directive the specular component is, which is similar to a generalized Lambertian pattern, that is, the greater the value of n the stronger the directivity of the specular component is, and θ_i is the incident angle of the incoming ray to the surface. Figure 3.9 shows an example of a radiation pattern generated using Phong's model, where r_d is taken to be 40% and $n = 30$. It can be observed from the figure that the specular component is radiated at a direction dependent on the incidence angle of the incoming ray.

3.6   Received Power from LOS Links

In an LOS link, optical power from a source can be received either by the receiver of the system or a small element on any reflecting surface, if the reflecting surface is assumed to be a summation of many small reflecting elements. In this case, the element acts as a point source after it has received optical power. The emitted power of the point source is the received power of the element multiplied by the reflecting coefficient of the surface. The emission pattern of the point source is assumed to be Lambertian and has been described in Section 3.5.1.

When an LOS link exists between a source and a receiver, the received optical power can be expressed by

$$h^{(0)}(t;\mathcal{S},\mathcal{R}) \approx \frac{n+1}{2\pi} P_S \cos^n(\theta)\, d\Omega\, \text{rect}\left(\frac{\psi}{\text{FOV}}\right)\delta\left(t-\left(\frac{d}{c}\right)\right) \qquad (3.6)$$

where $\mathcal{S}$ and $\mathcal{R}$ denote that the impulse response is defined for a specific configuration of sources and receivers, respectively; n is the Lambertian mode number of the source, P_S is the emitted power from the source; θ is the angle between the normal of the source $\hat{\mathbf{n}}_S$ and the direction of the emitted ray to the receiver; ψ is the angle between the normal of the receiver $\hat{\mathbf{n}}_R$ and the direction of the incoming ray from the source; FOV is the field-of-view of the receiver, defined as an angle such that only when ψ is less than FOV, the receiver detects light; d is the distance between the source and the receiver; c is the speed of light; $d\Omega$ is the solid angle subtended by the receiver's area A_R, that is, $d\Omega \approx \cos(\psi)\left(A_R/d^2\right), A_R \ll d^2$; and rect(.) is the rectangular function defined by $\text{rect}(x) = \begin{cases} 1 & \text{for } |x| \le 1 \\ 0 & \text{for } |x| > 1 \end{cases}$. Additionally, the point source $\mathcal{S}$ is further speci-

fied as $\mathcal{S} = \{\mathbf{r}_S, \hat{\mathbf{n}}_S, n\}$, where $\mathbf{r}_S$ is the position vector of the source, and the receiver $\mathcal{R}$ is further specified as $\mathcal{R} = \{\mathbf{r}_R, \hat{\mathbf{n}}_R, A_R, \text{FOV}\}$, where $\mathbf{r}_R$ is the position vector of the receiver. Figure 3.10 shows these parameters more clearly.

The unit vectors $\hat{\mathbf{n}}_S$ and $\hat{\mathbf{n}}_R$ should be specified in terms of conventions followed for room coordinates. The source and the receiver can have elevation and azimuth properties which have to be converted to unit vectors $\hat{\mathbf{n}}_S$ and $\hat{\mathbf{n}}_R$. Elevation of the source is the angle that $\hat{\mathbf{n}}_S$ makes with the xy plane, so if a source is directly pointed downward, elevation will be $-90°$, and if it is directly pointed upward, elevation will be $+90°$. Azimuth of the source is defined as the angle that the projection of $\hat{\mathbf{n}}_S$ on the xy plane makes with positive x-axis, with a sign such that positive y-axis has an azimuth of $+90°$. Hence, the conversions of elevation and azimuth angles of the source to $\hat{\mathbf{n}}_S$ can be given by

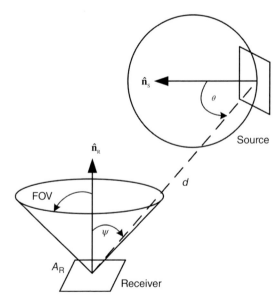

Figure 3.10   Source and receiver along with their related parameters.

$$\hat{\mathbf{n}}_{S\_x} = \cos(\text{elevation}) \times \cos(\text{azimuth})$$
$$\hat{\mathbf{n}}_{S\_y} = \cos(\text{elevation}) \times \sin(\text{azimuth})$$
$$\hat{\mathbf{n}}_{S\_z} = \sin(\text{elevation})$$

(3.7)

The definitions of elevation and azimuth of the receiver are similar to that of the source, except that $\hat{\mathbf{n}}_R$ is used in the definitions instead of $\hat{\mathbf{n}}_S$, and the calculations for $\hat{\mathbf{n}}_R$ for the conversions are exactly similar to (3.7).

3.7 Received Power from NLOS Links

Not all channels will have LOS link with the transmitter and the receiver. Regardless of whether an LOS link exists, which should have been determined by the previous step described, the next step is to calculate received power from reflections. This is the NLOS link. The total impulse response of the channel is the sum of responses from LOS and all reflections from NLOS links, given by

$$h(t;\mathcal{S},\mathcal{R}) = \sum_{k=0}^{\infty} h^{(k)}(t;\mathcal{S},\mathcal{R})$$

(3.8)

where $h^{(0)}$ is the LOS response that is calculated using (3.6) and $h^{(k)}$, $k>1$ are the responses of the kth reflections.

The major variations in modeling algorithms of NLOS links that are currently available in the literature can be broadly divided into two categories: deterministic and probabilistic methods. We will next describe the major algorithms and their variations to illustrate their usefulness in modeling NLOS OWC channels.

3.7.1 Barry's Algorithm

The first modeling algorithm for NLOS channels was reported by Gfeller and Bapst [5], where the idea of calculating the total power at the receiver by numerically integrating it over all the reflecting walls of the room was first presented. However, simulations reported using this method were limited to first reflections only. Impulse responses obtained by considering only the first reflections of the traveling photons can be used for rough approximations of the total power at the receiver, which may be sufficient for link budget analyses, but such impulse responses will not be adequate for precise calculation of delay spread and bandwidth of the channel. Barry *et al.* [9] first reported an algorithm that was able to calculate impulse responses considering multiple reflections. The total impulse response that includes all reflections in an indoor environment can be described by the following equation:

$$h^{(k)}(t;\mathcal{S},\mathcal{R}) = \int_S h^{(0)}\left(t;\mathcal{S},\left\{\mathbf{r},\hat{\mathbf{n}},\pi/2,d\mathbf{r}^2\right\}\right) \otimes h^{(k-1)}\left(t;\left\{\mathbf{r},\hat{\mathbf{n}},1\right\},\mathcal{R}\right)$$

(3.9)

where the integration is performed for all surfaces of the room and the operator is the convolution operator. The main idea to realize this integration is to perform a summation instead of integration, as follows:

$$h^{(k)}\left(t;\mathcal{S},\mathcal{R}\right) \approx \sum_{i=1}^{N} h^{(0)}\left(t;\mathcal{S},\left\{\mathbf{r}_i,\hat{\mathbf{n}}_i,\pi/2,d\mathbf{r}_i^2\right\}\right) \otimes h^{(k-1)}\left(t;\left\{\mathbf{r}_i,\hat{\mathbf{n}}_i,1\right\},\mathcal{R}\right) \qquad (3.10)$$

where the reflecting surfaces of the room, that is, walls, ceiling, floor, and furniture, are divided into N small reflecting elements. For the first reflections, each element is considered as a receiver that receives light in an LOS direction from the primary transmitter of the channel. Hence, the receivers have a FOV of $\pi/2$ as shown in (3.10). Then each element is considered as a separate transmitter, where it is considered as having Lambertian emission pattern with mode number $n=1$, and the received optical power at the primary receiver of the channel in LOS direction with the elements can be calculated. This gives the approximate contribution of first reflections to the total impulse response of the channel. It is approximate because the accuracy will get better as the reflecting elements get smaller, which also increases the computing time. For the second reflections, a different set of data is generated where, for each of the small reflecting elements being considered as a receiver, the rest of the elements are considered as transmitters, excluding the primary transmitter of the channel. Then, similar to the case of first reflections, the received optical power at the primary receiver of the channel in LOS direction with the small reflecting elements can be calculated. This gives the approximate contribution of second reflections to the total impulse response of the channel. More reflections can be considered exactly in the same way. This method is thus a recursive one and takes a considerable amount of computing time. Three reflections have been considered in Ref. [9] because of the amount of computing time needed, which may be sufficient for lower data rate communication systems. But as shown in Ref. [10], for high data rate communication systems, more orders of reflections should be considered and hence improvement over Barry's algorithm to reduce computing time was essential.

Improvements over Barry's method that have been reported in the literature hence primarily aim to decrease the computation time. The method in Ref. [11] is such an attempt where the simulations are not sliced into reflections but into time steps, and techniques such as storing some intermediate result in a disk file and reusing those results for later calculations leads to speed increase. One of the principal reasons that a large amount of computing time is required by Barry's method is the recursive nature of the algorithm. Hence, a big speed boost is to be expected if the algorithm is modified to make it an iterative one. This is precisely the modification reported in Ref. [12]. In Ref. [13], the iterative approach is further modified to include multiple transmitters and receivers. It should be mentioned that originally Barry's simulations in Ref. [9] did not include furniture or other reflecting materials than the walls, the ceiling, and the floor in the room, rather the simulations were considered for empty rooms only. References [12–14] include simulations considering furniture, thus showing some effects of shadowing. Including other reflecting materials in simulations is actually a straightforward extension of the algorithm and does not require any modification of the method. Simulations in Ref. [14] are done utilizing the unmodified version of Barry's algorithm. A small improvement is reported in Ref. [15] that describes a method of saving some calculations by disregarding reflecting elements as transmitters that are on the same surface as the element that is being considered as a receiver, since Lambertian pattern will result in zero intensity at directions in the same plane of the source. Next, we consider multiple-input multiple-output (MIMO) modeling method—an extension of Barry's algorithm.

3.7.2 MIMO Modeling Method

An important application in terms of multiple transmitters and receivers as a MIMO system is considered in Ref. [16], where intermediate calculations are saved in disk files and reused when a different transmitter or receiver position is assumed, thus avoiding the restart of the whole process of calculation by only considering the changed position. This allows generating a profile of a whole room where the receiver may be placed at many locations, and by relatively quick generation of impulse responses of all the receiver locations, delay spreads and received power contour plots can be obtained for the whole room.

In this model, the transfer function between a transmitter and a receiver is divided into four components. The first component represents the transfer function between a source and surface elements. The second component contains the transfer function between surface elements. The third has the transfer function from surface elements to a receiver. The last component accounts for direct response between a source and a receiver.

The surfaces of the room are made up of N neighboring elements of equal area, similar to Barry's algorithm. Elements are numbered sequentially, and each element is identified by an index. Diffusing spots are referred to as sources and the device generating them as transmitter. The number of reflections is counted from a source to a receiver.

3.7.2.1 Source Profile (F)

The first component in the model represents transfer function between source (diffusing spot) and surface elements. It is referred to as Source Profile and is modeled by a single-input multiple-output (SIMO) system with N outputs. The transfer function between a source and each of the surface elements is expressed by an entry in a vector M_S. Since surface elements 1 through N receive the signal directly, the transfer function f_{sk} between a source s and an element k is given by

$$f_{sk} = \rho_k \delta\left(t - \frac{d_{Ts}}{c}\right) \frac{\cos(\theta_{sk})\cos(\psi_{sk})A_R}{\pi d_{sk}^2} \delta\left(t - \frac{d_{sk}}{c}\right) \text{rect}\left(\frac{\theta_{sk}}{\pi/2}\right) \tag{3.11}$$

where ρ_k is the reflectivity of the element, the next term is added to account for the transfer function between a transmitter and a source, and the remaining terms are similar to the terms of (3.6) where A_R is the area of the element, d_{sk} is the distance between the source and the element, θ_{sk} is the angle between the normal of the source and the direction of the emitted ray to the element, ψ_{sk} is the angle between the normal of the element and the direction of the incoming ray from the source, and the FOV of an element is 90°.

For a single source s, the vector $\mathbf{F}_s$ is expressed as

$$\mathbf{F}_s = \begin{bmatrix} f_{s1} & \cdots & f_{sN} \end{bmatrix} \tag{3.12}$$

The expression can be extended to include more than a single source. In the case of a diffused link, where the transmitter illuminates the ceiling, the equivalent vector $\mathbf{F}_{eq}$ is defined as the equivalent of $n_x \times n_y$ sources and is given by

$$\mathbf{F}_{eq} = \sum_{i=1}^{n_x \times n_y} \mathbf{F}_i = \left[f_{eq1} \quad \cdots \quad f_{eqN} \right] \tag{3.13}$$

where $f_{eqj} = \sum_{i=1}^{n_x \times n_y} f_{ij}$.

3.7.2.2 Environment Matrix (Φ)

The second component consolidates dependence on indoor geometry, dimensions, and reflection coefficients. This component contains the transfer functions between any two reflecting elements. In matrix format, and considering up to n reflections, it is expressed as

$$\Phi_n = \begin{cases} \mathbf{I}_{N \times N} + \varphi + \varphi^2 + \varphi^3 + \cdots \varphi^{n-1}, & n \geq 2 \\ \mathbf{I}_{N \times N}, & n = 1 \end{cases} \tag{3.14}$$

where $\mathbf{I}_{N \times N}$ is the $N \times N$ identity matrix, and φ is given by

$$\varphi = \begin{bmatrix} \varphi_{11} & \cdots & \varphi_{1N} \\ \vdots & \ddots & \vdots \\ \varphi_{N1} & \cdots & \varphi_{NN} \end{bmatrix} \tag{3.15}$$

The entry φ_{ik} represents the transfer function between two elements i and k, and is given by

$$\varphi_{ik} = \begin{cases} 0, & i = k \\ \dfrac{\rho_i \cos\theta_{ik} \cos\psi_{ik} A_k}{\pi d_{ik}^2} \delta\left(t - \dfrac{d_{ik}}{c}\right) \mathrm{rect}\left(\dfrac{\theta_{ik}}{\pi/2}\right), & i \neq k \end{cases} \tag{3.16}$$

where the symbols carry their usual meanings similar to previous equations. The environment matrix is independent of transmitter and receiver. Once calculated, it can be used with any transmitter and receiver configuration.

3.7.2.3 Receiver Profile (G)

This component contains impulse response dependence on receiver parameters such as location and FOV. It contains transfer functions between a receiver and N surface elements. In vector form, it is expressed as

$$\mathbf{G}_r = \begin{bmatrix} g_{1r} \\ \vdots \\ g_{Nr} \end{bmatrix} \tag{3.17}$$

where the entry g_{ir} is given by

$$g_{ir} = \frac{\rho_i \cos\theta_{ir} \cos\psi_{ir} A_r}{\pi d_{ir}^2} \delta\left(t - \frac{d_{ir}}{c}\right) \mathrm{rect}\left(\frac{\theta_{ir}}{\mathrm{FOV}_r}\right) \quad (3.18)$$

where the symbols carry their usual meanings similar to previous equations.

3.7.2.4 Direct Response Vector (D)

When a source is within a receiver FOV, a direct response results. This response is expressed as

$$\mathbf{H}^{(0)} = \mathbf{DG}_r \quad (3.19)$$

where $\mathbf{D}$ is a $1 \times N$ vector given by

$$\mathbf{D} = \begin{bmatrix} d_1 & \cdots & d_N \end{bmatrix} \quad (3.20)$$

The entry d_i is equal to $(1/\rho_i)\delta\left(t - (d_{Ti}/c)\right)$ if an element i corresponds to a diffusing spot and 0, otherwise. d_{Ti} accounts for the delay between the transmitter and the source i. In a diffused configuration, there are $n_x \times n_y$ nonzero elements in $\mathbf{D}$.

3.7.2.5 Total Response (H)

The total impulse response $\mathbf{H}$ between a source s and a receiver when n reflections are considered can be expressed as

$$\mathbf{H} = \sum_{i=0}^{n} \mathbf{H}^{(i)}$$
$$= \mathbf{DG}_r + \mathbf{F}_s \Phi_n \mathbf{G}_r \quad (3.21)$$

If vector Ψ_n contains N entries defined as

$$\Psi_n = \mathbf{D} + \mathbf{F}_s \Phi_n \quad (3.22)$$

$\mathbf{H}$ can be written as

$$\mathbf{H} = \Psi_n \mathbf{G}_r \quad (3.23)$$

The vector Ψ_n contains the signal components seen by each surface element. Multiplying by $\mathbf{G}_r$, the signal is shifted to account for the delay between elements and receiver and is multiplied by a factor that depends on the path between each element and receiver. The expression for $\mathbf{H}$ in (3.23) readily applies to a diffused link if $\mathbf{F}$ is substituted by $\mathbf{F}_{eq}$.

One of the advantages attained by this model is highlighted in (3.23). In analyzing an indoor room environment, we are often interested in the impulse response for many receiver locations within a room. Room parameters do not change, nor do parameters of transmitter, only parameters of receiver change. Using (3.23) to calculate a new impulse response requires calculating the new value of $\mathbf{G}_r$ and multiplying by Ψ_n, which is already calculated. The time required for calculating Ψ_n is comparable to that required to calculate a single impulse response. Once Ψ_n is calculated and stored, however, the time required to calculate a new impulse response is reduced to calculating the multiplication of two matrices, which takes a very short time. This model saves time even when a single impulse response is calculated. By calculating receiver vector $\mathbf{G}_r$ first, elimination of any unnecessary calculation is possible. This is especially true when the receiver FOV is small, since the nonzero entries in $\mathbf{G}_r$ are a small fraction of total entries. When the kth entry in $\mathbf{G}_r$ is zero, the corresponding kth column in Ψ_n does not affect the calculation since it is multiplied by zero. Therefore, for n nonzero entries in $\mathbf{G}_r$, only corresponding n columns in Ψ_n have to be calculated. Figure 3.11 shows the MIMO structure of the model.

3.7.3 Modified Monte Carlo Algorithm and Variations

An alternative method for calculating impulse responses using Monte Carlo ray-tracing techniques is shown in Ref. [17]. The Lambertian pattern is considered as a probability density function (pdf), and rays are traced carrying equal optical power from the source by generating their directions using this pdf. As the rays hit reflecting surfaces, new rays are generated by the same pdf while their power is reduced by the reflectivities of those surfaces. Thus, by generating random directions using the pdf, rays are traced throughout the room, and as they reach the receiver, an impulse response can be found. There is no limitation on the number of

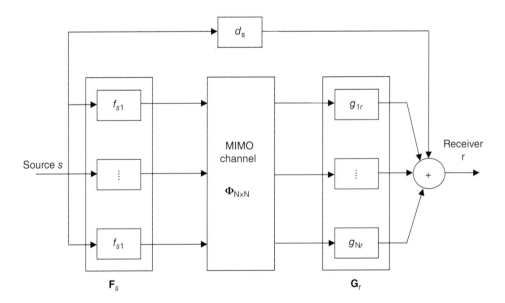

Figure 3.11   MIMO modeling method.

reflections that can be considered in this method. However, the probability that the rays will reach the receiver is not high and so a very large number of rays need to be traced. To remedy this, a modified version of the method, called the Modified Monte Carlo (MMC) algorithm has been introduced in Ref. [18–20], where each reflection of the rays is used to calculate an LOS contribution to the receiver from the reflecting point, thus utilizing each ray multiple times instead of only once. This leads to a lower number of generated rays from the source to calculate an impulse response. The algorithm is purely sequential and iterative, and hence this method is very fast. However, the impulse response obtained in this way contains some variance because of the random nature of the direction of the rays. This variance decreases as the number of rays is increased. An error analysis of this method of calculating impulse responses appears in Ref. [21]. It should be noted that there is provision in the algorithm to consider specular reflections off the surfaces too instead of only Lambertian; that is, deterministic directions can be considered as required instead of random directions; however, this just determines whether different reflecting coefficients of the surfaces will be considered or not depending on the nature of reflection and does not change the basic workflow of the algorithm. A versatile tool is reported in Ref. [22] that employs the MMC algorithm for calculating impulse responses for any environment. The environment description is specified as a 3D computer aided design (CAD) model, and hence the tool supports environment geometries having any shape, size, area, and number of obstacles including their material properties such as reflectivity. The MMC algorithm is especially suitable for parallelization leading to further reduction of the total execution time. Results are reported in Ref. [22] that show computational speed-up proportional to the number of processors used in the simulation. Section 3.7.4 describes an algorithm that incorporates the MMC method as part of its workflow. We will describe the MMC algorithm in more detail in Section 3.7.4.

A slightly different approach to the MMC algorithm is reported in Ref. [23] where the power of the rays is not reduced after a reflection, rather whether the rays will be propagated further or not after hitting a reflecting surface is determined by the reflecting coefficients. Thus the number of rays decreases as more reflections are considered, and hence it is a faster version of the MMC method. However, even more variance is introduced by this procedure, and as reported in Ref. [24], the MMC algorithm generates impulse responses with a smaller amount of estimated errors compared to it. Another algorithm similar to these is reported in Ref. [25], where individual photons are traced, and their propagations after hitting a surface are determined by the bidirectional reflectance distribution function of that surface, which is a general function describing reflectance of a surface that can be simplified to other forms, such as Lambertian pattern for perfectly diffuse surfaces.

3.7.4   Combined Deterministic and MMC Algorithm

The best accuracy can be achieved by Barry's method and its modifications depending on how small the reflecting elements are, while the best speed can be achieved by the MMC algorithm. In this section, a method is described where both of these processes are combined. The surfaces are divided into small reflecting elements, and the contribution of first reflections to the total impulse response is found using these elements as is done by Barry's method. However, the contributions of the second and remaining reflections are calculated by the MMC method, where each of those elements is considered as a source. In the original MMC simulations, a very large number of rays are needed to be generated from the light source to obtain an impulse response with sufficiently acceptable variance. In this method, only a couple of rays are required to be generated from each reflecting element since there are a large number of elements in the room. Depending on configurations, the contribution of first reflections to the total impulse response may contain the most significant amount of reflected power, and since this is calculated by Barry's method, very good accuracy can be achieved. The contributions of the remaining reflections to the total impulse response contain a lesser amount of power, depending on configurations, and hence some variance in those contributions introduced by using the MMC method is acceptable. Hence, the best features of both processes can be utilized. The algorithm is sequential and iterative and hence very fast. It can be implemented for multi-core systems, that is, the algorithm is parallelizable. It is also possible to save the results of the first reflection contributions to a disk file and later reuse them when considering a new location of the receiver. This will allow generating a profile of a whole room very quickly, where the receiver may be placed at many locations. Since the algorithm contains both deterministic elements as well as a modified version of the Monte Carlo method, it is known as the Combined Deterministic and Modified Monte Carlo (CDMMC) method of calculating impulse responses.

3.7.4.1   Contribution of First Reflections to the Total Impulse Response

As the first step of the CDMMC method, the contribution of the LOS response from the source to the receiver, if LOS link exists, is calculated by using (3.6).

The next step is to calculate the contribution of first reflections to the total impulse response. This is done by first dividing each surface of the room into small reflecting elements. The width and the height of the elements, that is, the area of the elements, will determine the errors of the contribution of first reflections compared to the contribution when there is no discretization involved, and also will determine the variance of contributions of the second and remaining reflections to the total impulse response. A reasonable value of this area is $1 \times 10^{-4} \, \text{m}^2$. The elements are assumed to be squares. The position of each element is determined by the center of the square (intersection of the diagonals). Each of the elements is first considered to be a receiver, hence for each of them $\hat{\mathbf{n}}_R$ is specified. The FOV of the elements is assumed to be $90°$. Thus, by using (3.6), it is possible to calculate the received power at each element from the source. The time required for light to travel from the source to the elements is recorded for each of them.

Next, each element is considered as a point source, and the received power at the receiver from each of them is calculated by using (3.6) again. But in this case, $\hat{\mathbf{n}}_S$ is the same unit vector $\hat{\mathbf{n}}_R$ that was considered in the previous step. The optical power, P_S, of the reflecting elements is obtained by multiplying their received power calculated in the last step by the reflecting coefficient of the surface to which each of them belongs. The Lambertian mode number n of the elements when they act as point sources is taken to be 1, which is a very close approximation. Thus, by using (3.6), the LOS power at the receiver is calculated for each reflecting element. Also, the time required for light to travel from each of the elements to the receiver is recorded, which is added to the time previously calculated for denoting time required for light to travel from the source to the elements. Hence, the total time required for light to travel from the source to each small reflecting element and then to the receiver is found. At these time values, in an array denoting $h^{(1)}(t)$, the power contributions are added from each element, and thus the contribution of first reflections to the total impulse response is obtained.

3.7.4.2 Contributions of Second and Subsequent Reflections to the Total Impulse Response

Contributions of second and subsequent reflections to the total impulse response are calculated by the MMC method. The basic principles of the method have been introduced in Ref. [20]. The difference between [20] and this algorithm is that here each of the small reflecting elements considered in the previous step is used as a source to generate rays instead of using the original light source.

In Monte Carlo method, the rays are generated probabilistically from the sources in the following way. It is observed that (3.1) can be interpreted as a probability density function (pdf) if P_S is equal to 1 W, since the result of the total volume integration of (3.1) is P_S. Thus, (3.1) can be rewritten as a function of θ keeping $P_S = 1$ as

$$f(\theta,\varphi) = \frac{n+1}{2\pi} \cos^n(\theta)d\varphi d\theta \tag{3.24}$$

Equation (3.24) can be considered as a pdf for the random variable Θ. It should be noted that it is not necessary to consider φ as a value of a random variable since Lambertian pattern is symmetric around the normal of the surface of the source. Volume integration of (3.24) through all possible values of θ and φ yield the result unity, and thus the assumption of (3.24) to be a pdf is correct. Hence, the cumulative distribution function (cdf) of Θ, that is, the probability that Θ is less than or equal to θ, can be obtained in the following way:

$$P(\Theta \le \theta) = \int_{\varphi=0}^{2\pi} \int_{\beta=0}^{\theta} f(\beta,\varphi)\sin(\beta)d\varphi d\beta \tag{3.25}$$
$$= 1 - \cos^{n+1}(\theta)$$

The sine term in the integration of (3.25) appears from the Jacobian due to change in variables from Cartesian to spherical coordinates.

Now, another random variable Z from the random variable Θ can be obtained by

$$Z = \cos\Theta \tag{3.26}$$

Z represents the coordinates in the z-axis. The cdf of Z can be calculated by the following steps:

$$\begin{aligned} P(Z \le z) &= P(\cos\Theta \le z) \\ &= P(\Theta \ge \cos^{-1} z) \\ &= 1 - P(\Theta \le \cos^{-1} z) \\ &= (\cos\cos^{-1} z)^{n+1} \\ &= z^{n+1} \end{aligned} \tag{3.27}$$

The goal here is to obtain a random unit vector that will represent the coordinate system for a randomly generated ray as shown in Figure 3.12. The unit vector's xy plane is the surface plane of the elements in the CDMMC algorithm.

The z-component of the unit vector has a cdf as in (3.27). The x- and the y-components of the unit vector need to be generated. To obtain the values of x-, y-, and z-components, two random variables U and V are generated having uniform distribution in the range [0, 1]. Samples of random variable Z (values of z-components of the unit vector) can be generated by using inverse transform sampling method from its cdf obtained in (3.27) and samples of the random variable U. x- and y-components can be generated by using the values of z-components and samples of the random variable V. The steps are as follows:

$$
\begin{aligned}
z &= \sqrt[n+1]{u} \\
x &= \sqrt{1 - z^2}\, \cos(2\pi v) \\
y &= \sqrt{1 - z^2}\, \sin(2\pi v)
\end{aligned}
\tag{3.28}
$$

Thus, the x-, y-, and z-components of the unit vector are obtained. Each unit vector represents the direction of a ray that is emitted from the reflecting elements. It should be noted here that these components are based on a coordinate system whose xy plane is considered to be the element's surface. Since the elements can belong to any surface of the room, these coordinates are not the same as room coordinates, and thus have to be converted to room coordinates to map the path of the ray. Assuming the elements are located on the walls, the ceiling, and the floor only, or on surfaces that are perpendicular to the walls, we will detail the transformation of element coordinates to room coordinates at the end of this section. If the elements in question belong to a surface that is not perpendicular to the walls, but is slanted, the transformation process is different and we will also describe it at the end of this section.

Each of the elements from the step of calculating the contribution of first reflections to the total impulse response is considered to generate random rays. Suppose a particular reflecting element receives a certain amount of optical power, say, $P_{elem\_receive}$, in the first step. When it is acting as a point source, its emitted power is $P_{elem\_emit} = P_{elem\_receive} \times \rho_{surface}$, where $\rho_{surface}$ is the reflecting coefficient of the surface to which the element belongs. Now,

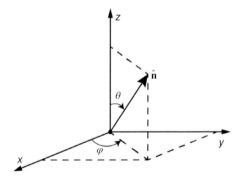

Figure 3.12   Coordinate system for a random ray (denoted by a random unit vector).

for the Monte Carlo method to work properly, usually a very large number of rays is required to be generated as is shown in Ref. [20], ranging to about half a million or more. However, in Ref. [20] the rays are generated from the single light source of the system, and here rays are generated from a large number of reflecting elements acting as sources. From simulations we found that only about 10 rays generated from each element is sufficient to calculate contributions of second and subsequent reflections to the total impulse response with acceptable variance, assuming a normal room size and an element area of $1 \times 10^{-4}\,\mathrm{m}^2$. For large halls/rooms and for element areas larger than the value mentioned here, more than 10 rays should be generated. So, if $N_{ray} = 10$, optical power assigned to each of these rays is $P_{ray} = P_{elem\_emit} / N_{ray}$.

Each of the generated rays propagates through the room and hits a surface. A fast and efficient process of calculating this impact point is also detailed at the end of this section. The time required for the ray to traverse from the element to the impact point is recorded. A new ray with a random direction is generated from this impact point using the steps mentioned in (3.28). The optical power carried by this new ray is equal to the optical power of the previous incoming ray to the impact point multiplied by the reflecting coefficient of the surface to which the impact point belongs. This new ray is again propagated through the room, a new impact point is found, time required for the new ray to travel from its origin to the new impact point is recorded, and another ray is similarly generated, and the process continues. At some point, a ray may reach the receiver of the system and (3.6) can be applied to find out the received power, along with the total time required for all these reflections to reach the receiver.

As it is readily understood, the probability that rays will reach the receiver is quite low. Hence in Ref. [20], an MMC method is introduced, which utilizes each ray to contribute to the power received by the receiver. A similar method is followed here too. When the first impact point is calculated, (3.6) is readily applied to get the power received by the receiver where the impact point is the source with Lambertian mode number $n = 1$ as mentioned earlier, and P_S is found by multiplying the power of the incoming ray to the impact point by the reflecting coefficient of the surface to which the impact point belongs. The corresponding time required for light to travel from the impact point to the receiver is calculated and added to the previous cumulative time recorded. Hence, in an array denoting $h^{(2)}(t)$, the calculated received powers are added for all rays generated from all reflecting elements at their corresponding cumulative recorded times, and thus the contribution of second reflections to the total impulse response is obtained.

This process is continued in a similar way. We have mentioned that from the first impact point of a ray, a new ray with randomly generated direction is propagated through the room and a second impact point is found, along with a new cumulative time. Similar to the method discussed above, power received by the receiver from this second impact point by using (3.6) is calculated, along with the added time required for light to travel the total distance from the source to the element, then to the first impact point, and then second impact point, and finally to the receiver. Repeating for all rays and all elements, the contribution of third reflections to the total impulse response is obtained.

There is no limit whatsoever in the algorithm on how many reflections can be considered. Depending on the requirement of how long the tail of the total impulse response needs to be calculated, the number of reflections can be varied. But at least three reflections should be considered as they carry the most significant amount of power [9, 20].

Transformation of Element Coordinates to Room Coordinates for Surfaces Perpendicular to Walls

In the Monte Carlo section of the CDMMC algorithm, directions of the rays are generated probabilistically. This direction is indicated by a unit vector. However, the coordinate system that the unit vector is based on is the local coordinate system of the reflecting surface which is, in most cases, not the same as the coordinate system followed for the room as shown in Figure 3.4. In Figure 3.13, the coordinate axes on all six surfaces of an empty room are shown, where the origins of the individual coordinate systems are not important as the unit vectors will point out the directions only. However, these coordinate axes have to be rotated to resemble the directions of coordinate axes in Figure 3.4 so that the unit vectors can point out directions in room coordinates. This is done by pre-multiplying the directions of the unit vectors with appropriate rotation matrices.

The rotation matrices are defined as follows:

$$\mathbf{R}_x = \begin{bmatrix} 1 & 0 & 0 \\ 0 & \cos(\theta_x) & \sin(\theta_x) \\ 0 & -\sin(\theta_x) & \cos(\theta_x) \end{bmatrix} \tag{3.29}$$

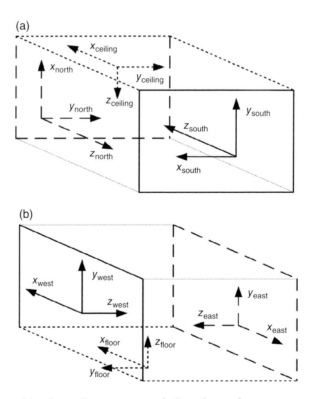

Figure 3.13 Axes of local coordinate systems of all surfaces of an empty room, neglecting the individual origins of the systems: (a) north wall, south wall, and the ceiling; (b) east wall, west wall, and the floor.

$$\mathbf{R}_y = \begin{bmatrix} \cos(\theta_y) & 0 & -\sin(\theta_y) \\ 0 & 1 & 0 \\ \sin(\theta_y) & 0 & \cos(\theta_y) \end{bmatrix} \tag{3.30}$$

$$\mathbf{R}_z = \begin{bmatrix} \cos(\theta_z) & \sin(\theta_z) & 0 \\ -\sin(\theta_z) & \cos(\theta_z) & 0 \\ 0 & 0 & 1 \end{bmatrix} \tag{3.31}$$

where the subscript in the matrices indicate which axis remains fixed during the rotation, that is, the axis about which the rotation occurs, and the angles represent anti-clockwise rotation of the other two axes when the angle is positive and clockwise rotation when the angle is negative. For example, the rotation matrix $\mathbf{R}_x$, when pre-multiplied with a unit vector, will rotate the coordinate system in such a way that the rotation will be about the x-axis of the local coordinate system, and θ_x represents the angle at which the y-axis rotates anti-clockwise toward z-axis when it is positive and clockwise when it is negative. Thus, if a unit vector in the local coordinate system of the surface is represented by $\hat{\mathbf{d}}_{local} = \begin{bmatrix} x_{local} & y_{local} & z_{local} \end{bmatrix}^T$, we can convert it to room coordinates by applying the following:

$$\hat{\mathbf{d}}_{room} = \begin{bmatrix} x_{room} \\ y_{room} \\ z_{room} \end{bmatrix} = \mathbf{R}_x \mathbf{R}_y \mathbf{R}_z \hat{\mathbf{d}}_{local}$$

$$\Rightarrow \begin{cases} x_{room} = x_{local} \cos(\theta_y)\cos(\theta_z) + y_{local} \cos(\theta_y)\sin(\theta_z) - z_{local} \sin(\theta_y) \\ y_{room} = -x_{local} \cdot (\cos(\theta_x)\sin(\theta_z) - \sin(\theta_x)\sin(\theta_y)\cos(\theta_z)) \\ \qquad + y_{local} \cdot (\cos(\theta_x)\cos(\theta_z) + \sin(\theta_x)\sin(\theta_y)\sin(\theta_z)) \\ \qquad + z_{local} \sin(\theta_x)\cos(\theta_y) \\ z_{room} = x_{local} \cdot (\sin(\theta_x)\sin(\theta_z) + \cos(\theta_x)\sin(\theta_y)\cos(\theta_z)) \\ \qquad - y_{local} \cdot (\sin(\theta_x)\cos(\theta_z) - \cos(\theta_x)\sin(\theta_y)\sin(\theta_z)) \\ \qquad + z_{local} \cos(\theta_x)\cos(\theta_y) \end{cases} \tag{3.32}$$

To rotate the directions of the local coordinate axes of the surfaces as shown in Figure 3.13 to align with the directions of the room coordinate axes as shown in Figure 3.4, from the definitions of the rotation matrices and angles discussed above, the angles given in Table 3.1 can be obtained. By applying (3.32) with these angles, the direction of any unit vector in reference to a local coordinate axis of a surface can be converted to the direction in reference to the room coordinates. Though the equations in (3.32) look involved, it can be observed that replacing the values of angles from Table 3.1 in (3.32) leads to very simple equations for each surface. Also it should be noted that though only six surfaces are shown

Table 3.1 Rotation angles of different surfaces

Surfaces	Angles		
	θ_x	θ_y	θ_z
North	$-\pi/2$	$\pi/2$	0
South	$-\pi/2$	$-\pi/2$	0
East	$\pi/2$	0	$-\pi$
West	$-\pi/2$	0	0
Ceiling	$-\pi$	0	0
Floor	0	0	0

here, any other reflecting surfaces can be treated in a similar way if they are perpendicular to the room walls.

Transformation of Element Coordinates to Room Coordinates for Slanted Surfaces

In this section, we discuss the method of converting the direction of any unit vector in reference to a local coordinate axis of a surface to the direction in reference to the room coordinates when the surface in question is not perpendicular to any walls, the ceiling, or the floor. In this case, the surface is slanted and perpendicular to its local xy plane, that is, the direction to which the surface is facing needs to be defined. We define it in terms of its azimuth, φ, and elevation, θ, and they are shown in Figure 3.14. The convention we follow here in order to obtain the local xy plane of the surface is as follows. First, the x-axis of the room coordinate system has to be rotated by the azimuth angle, φ, about the z-axis of the room coordinate system, where a positive angle denotes anti-clockwise rotation and a negative angle denotes clockwise rotation. The rotated y-axis is now termed as the local y-axis of the surface. Keeping the y-axis fixed at its position, another rotation is performed anti-clockwise to obtain the elevation angle, θ, which defines the angle between the local z-axis of the surface and the xy plane of the room coordinate system. That means, the rotation is performed anti-clockwise about the local y-axis by an angle of $90° - \theta$. The rotated x- and z-axes now indicate the local x- and z-axes of the surface, respectively.

With the conventions and definitions mentioned, the rotation matrix can be defined now to convert the direction of any unit vector in reference to a local coordinate axis of a slanted surface to the direction in reference to the room coordinates. It should be noted that this is simply a change of bases in terms of linear algebra, where the room coordinate system forms the standard basis, and the local coordinate system forms a different basis and the direction of the unit vector is given in terms of the latter basis. Hence we have

$$\mathbf{I}_{3\times3}\hat{\mathbf{d}}_{room} = \begin{bmatrix} x_{room} \\ y_{room} \\ z_{room} \end{bmatrix} = \mathbf{A}_{local,3\times3}\hat{\mathbf{d}}_{local} = \begin{bmatrix} \mathbf{x}_{local,3\times1} & \mathbf{y}_{local,3\times1} & \mathbf{z}_{local,3\times1} \end{bmatrix} \begin{bmatrix} x_{local} \\ y_{local} \\ z_{local} \end{bmatrix} \qquad (3.33)$$

where $\mathbf{I}_{3\times3}$ is the identity matrix representing the standard basis and $\mathbf{A}_{local,3\times3}$ is the basis representing the local coordinate system of the surface. The columns of this matrix represent the axes of the coordinate system. We can determine these column vectors graphically from

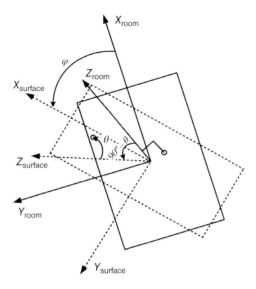

Figure 3.14   Azimuth and elevation of a slanted surface.

Figure 3.14 following the conventions of azimuth and elevation as stated earlier and obtain the following relations:

$$
\hat{\mathbf{d}}_{room} =
\begin{bmatrix} x_{room} \\ y_{room} \\ z_{room} \end{bmatrix}
= \mathbf{A}_{local,\,3\times3}\,\hat{\mathbf{d}}_{local} =
\begin{bmatrix}
\sin(\theta)\cos(\varphi) & -\sin(\varphi) & \cos(\theta)\cos(\varphi) \\
\sin(\theta)\sin(\varphi) & \cos(\varphi) & \cos(\theta)\sin(\varphi) \\
-\cos(\theta) & 0 & \sin(\theta)
\end{bmatrix}
\begin{bmatrix} x_{local} \\ y_{local} \\ z_{local} \end{bmatrix}
$$

(3.34)

$$
\Rightarrow
\begin{cases}
x_{room} = x_{local}\sin(\theta)\cos(\varphi) - y_{local}\sin(\varphi) + z_{local}\cos(\theta)\cos(\varphi) \\
y_{room} = x_{local}\sin(\theta)\sin(\varphi) + y_{local}\cos(\varphi) + z_{local}\cos(\theta)\sin(\varphi) \\
z_{room} = -x_{local}\cos(\theta) + z_{local}\sin(\theta)
\end{cases}
$$

It is possible to use (3.34) as a general method in the case of perpendicular surfaces too instead of using (3.32). However, the conventions of local axes shown in Figure 3.13 have to be changed to properly use the definitions of azimuth and elevation angles followed in (3.34).

Detection of an Impact Point of a Ray and a Surface
We show here a procedure to detect the impact point as a generated ray is propagated through the room. The impact point can be located on any reflecting surface. The procedure here is a step-by-step process, where first it is checked if the impact point is located on either the north or the south wall. If not, the check is continued for the west and the east walls. If not, the check is continued for the ceiling and the floor. The procedure can be extended for any number of surfaces as necessary.

For example, we show here first how a check for the north wall is done. We know the position vector of the point source from where the rays are generated. We need to know the position vector of the impact point. This can be expressed by the following:

$$
\mathbf{r}_{impact} = \mathbf{r}_{source} + k\hat{\mathbf{n}}_{ray}
$$

(3.35)

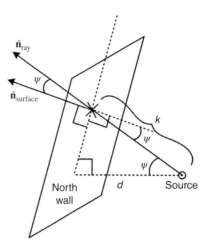

Figure 3.15   Geometry showing the method to determine the coordinates of the impact point.

where the unit vector $\hat{\mathbf{n}}_{ray}$ indicates the direction of the ray and k is the distance between the point source and the impact point. In Figure 3.15, the geometry for determining the distance k is shown.
$\hat{\mathbf{n}}_{surface}$ is the unit vector perpendicular on the surface we are checking. Since we are concerned with the north wall now, $\hat{\mathbf{n}}_{surface} = \left(1\hat{\mathbf{i}}, 0\hat{\mathbf{j}}, 0\hat{\mathbf{k}}\right)$, where the symbols carry their usual meanings. We also have $\hat{\mathbf{n}}_{ray} = \left(x_{ray}\hat{\mathbf{i}}, y_{ray}\hat{\mathbf{j}}, z_{ray}\hat{\mathbf{k}}\right)$. ψ is the angle between these two unit vectors. Then from their dot products we have,

$$\hat{\mathbf{n}}_{ray} \cdot \hat{\mathbf{n}}_{surface} = \left|\hat{\mathbf{n}}_{ray}\right|\left|\hat{\mathbf{n}}_{surface}\right|\cos(\psi)$$

$$\Rightarrow x_{ray} = 1 \times 1 \times \frac{d}{k} \tag{3.36}$$

$$\Rightarrow k = \frac{d}{x_{ray}}$$

where d is the perpendicular distance from the north wall to the point source, which is known from the position of the point source and the room length. Equation (3.35) is applied as the next step and the obtained coordinates are checked to see whether they fall into the boundaries of the north wall. If not, the next wall is checked using a similar procedure and the process continues.

3.7.4.3   Summary of the Steps of the CDMMC Algorithm

The CDMMC algorithm can be summarized as having the following steps:

Step 1.   Calculate the normal unit vectors of the source and the receiver with respect to their surface from their azimuth and elevation, if given.

Step 2.   Calculate LOS response from the source to the receiver.

Step 3.   Divide all surfaces of the room into small reflecting elements and record their normal unit vectors with respect to their surface.

Step 4.   Calculate LOS response from the source to all small reflecting elements. Keep track of time required for light to travel from the source to each element, that is, $t_{s\_elem} = d_{s\_elem} / c$, where d_{s_elem} is the distance from the source to an element and c is the speed of light.

Step 5.   Consider each element as a point source having Lambertian mode number $n = 1$. Power emitted by each element is $P_{elem\_emit} = P_{elem\_receive} \times \rho_{surface}$, where $\rho_{surface}$ is the reflecting coefficient of the surface to which the element belongs.

Step 6.   Calculate LOS response from each reflecting element to the receiver. Keep track of time, $t_{elem\_r} = d_{elem\_r} / c$, where d_{elem_r} is the distance from an element to the receiver.

Step 7.   Add all responses at times $t_{elem\_r} + t_{s\_elem}$. This will give the contribution of first reflections to the total impulse response.

Step 8.   Generate rays having random directions using Lambertian pattern as a pdf from each small reflecting element acting as a source. If number of rays generated from each element is N_{ray}, each ray will carry power $P_{ray} = P_{elem\_emit} / N_{ray}$.

Step 9.   Propagate each ray until a surface is hit. Find the impact point. Keep track of time, $t_{elem\_input} = d_{elem\_impact} / c$, where d_{elem_impact} is the distance from the element to the impact point.

Step 10.   From the impact point, calculate LOS response to the receiver. Keep track of time, $t_{impact\_r} = d_{impact\_r} / c$, where d_{impact_r} is the distance from the impact point to the receiver.

Step 11.   Add all responses at times $t_{impact\_r} + t_{elem\_impact} + t_{s\_elem}$. This will give the contribution of second reflections to the total impulse response.

Step 12.   Generate another ray from the impact point having power $P_{ray} = P_{inco\_ray} \times \rho_{surface}$, where P_{inco_ray} is the power of the incoming ray that hits this surface at the impact point and $\rho_{surface}$ is the reflecting coefficient of the surface to which the impact point belongs.

Step 13.   Propagate the ray until a surface is hit. Find the impact point. Keep track of time, $t_{impact\_impact} = d_{impact\_impact} / c$, where d_{impact_impact} is the distance between the two impact points.

Step 14.   From the new impact point, calculate LOS response to the receiver. Keep track of time, $t_{impact\_r} = d_{impact\_r} / c$, where d_{impact_r} is the distance from the new impact point to the receiver.

Step 15.   Add all responses at times $t_{impact\_r} + t_{impact\_impact} + t_{elem\_impact} + t_{s\_elem}$. This will give the contribution of third reflections to the total impulse response.

Step 16.   Repeat steps 12–15 to calculate contributions from as many reflections to the total impulse response as required.

3.7.4.4   Comparison of Impulse Responses Calculated by Different Algorithms

In this section we will show results of simulations of impulse responses of three room configurations given in Table 3.2. These room configurations are the same as reported in Ref. [9]. We have selected these as our test configurations since [9] has reported experimental results of impulse responses of these configurations too. The experimental results and the simulations

Table 3.2 Room configurations and parameters of the source and the receiver for simulations

	Room and surfaces	Source	Receiver
Configuration A	Room length: 7.5 m	Lambertian mode: 1	Area: $1 \times 10^{-4} m^2$
	Room width: 5.5 m	x: 2.0 m	x: 6.6 m
	Room height: 3.5 m	y: 4.0 m	y: 2.8 m
	Reflecting coefficients:	z: 3.3 m	z: 0.8 m
	ρ_{north}: 0.30	Elevation: −90°	Elevation: +90°
	ρ_{south}: 0.56	Azimuth: 0°	Azimuth: 0°
	ρ_{east}: 0.30		FOV: 70°
	ρ_{west}: 0.12		
	$\rho_{ceiling}$: 0.69		
	ρ_{floor}: 0.09		
Configuration B	Room length: 7.5 m	Lambertian mode: 1	Area: $1 \times 10^{-4} m^2$
	Room width: 5.5 m	x: 5.0 m	x: 2.0 m
	Room height: 3.5 m	y: 1.0 m	y: 4.0 m
	Reflecting coefficients:	z: 3.3 m	z: 0.8 m
	ρ_{north}: 0.58	Elevation: −70°	Elevation: +90°
	ρ_{south}: 0.56	Azimuth: 10°	Azimuth: 0°
	ρ_{east}: 0.30		FOV: 70°
	ρ_{west}: 0.12		
	$\rho_{ceiling}$: 0.69		
	ρ_{floor}: 0.09		
Configuration C	Room length: 7.5 m	Lambertian mode: 1	Area: $1 \times 10^{-4} m^2$
	Room width: 5.5 m	x: 3.75 m	x: 6.0 m
	Room height: 3.5 m	y: 2.75 m	y: 0.8 m
	Reflecting coefficients:	z: 1.0 m	z: 0.8 m
	ρ_{north}: 0.58	Elevation: +90°	Elevation: +90°
	ρ_{south}: 0.56	Azimuth: 0°	Azimuth: 0°
	ρ_{east}: 0.30		FOV: 70°
	ρ_{west}: 0.12		
	$\rho_{ceiling}$: 0.69		
	ρ_{floor}: 0.09		

reported in Ref. [9] match closely with each other and hence we have chosen to compare the calculated impulse responses here with those reported in Ref. [9], as the latter can be assumed to be completely correct.

As an example, the shape of the impulse response of configuration A is shown in Figure 3.16 to point out that the impulse response calculated by the CDMMC algorithm matches very closely with that calculated by Barry's algorithm [9] and the MMC method [20]. Configuration A has an LOS response between the source and the receiver. This has been shown in the figure. Since in the CDMMC algorithm, the contribution of first reflections to the total impulse response has been calculated using exactly the similar process as Barry's algorithm, that is, deterministically, it is without variance. However, the contributions of the remaining reflections to the total impulse response are generated in a similar process as the MMC method, and so they will not be as temporally smooth as the contribution of first reflections. This variance can be seen in the impulse responses generated by the MMC and the CDMMC methods as

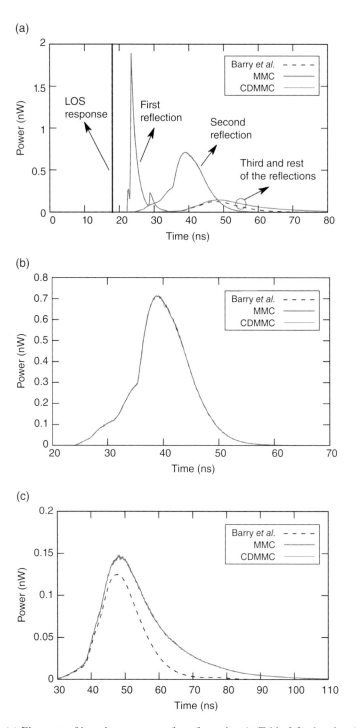

Figure 3.16 (a) Elements of impulse response of configuration A (Table 3.2) showing the LOS component, contributions of first, second, third, and remaining reflections to the total impulse response separately as calculated by Barry's algorithm, the MMC, and the CDMMC method, (b) a zoomed-in version of the contribution of second reflections to the total impulse response, and (c) a zoomed-in version of the contributions of third and remaining reflections to the total impulse response.

shown in Figure 3.16b and c that show the contribution of second reflections and contributions of third and remaining reflections to the total impulse response, respectively. The impulse response calculated by Barry's algorithm as shown in Figure 3.16 here contains only three reflections while the impulse responses calculated by the MMC and the CDMMC methods contain ten reflections. Hence, in Figure 3.16a and c, we observe higher power in the impulse responses that are related to the contributions of third and remaining reflections to the total impulse response. From Figure 3.16, it can be concluded that impulse responses calculated by all three algorithms are very similar in shape to each other.

3.7.4.5   Error Analysis

As noted in Section 3.7.4.4, the CDMMC method introduces some variance when calculating contributions of second and remaining reflections to the total impulse response since this part of the algorithm employs the Monte Carlo method. Though all three algorithms, namely Barry's algorithm, the MMC, and the CDMMC algorithms, calculate very similar impulse responses, it is important to analyze the nature of this variance to understand better their strengths and weaknesses.

The MMC algorithm introduces variance as it employs the Monte Carlo method. In Ref. [21], the nature of this variance and estimated error has been analyzed in detail. It is useful to analyze the estimated error in terms of relative estimated error, which is indicated by the ratio of the square root of the variance of the received power in a small time interval, or in other words, one standard deviation of the received power in a small time interval to the total power received in that time interval. From Ref. [21], this quantity is given by

$$\text{rel err } (P_k) = \frac{\sum_{i=1}^{N_k} p_{i,k}^2 - \frac{1}{N}\left(\sum_{i=1}^{N_k} p_{i,k}\right)}{P_k} \tag{3.37}$$

where N_k denotes the number of rays reaching the receiver in the kth time interval, $p_{i,k}$ is the amount of power contributed to the receiver by the ith ray in that interval, N is the total number of rays in flight in the interval which is equal to the number of rays emitted from the source if the time interval is small, and P_k is the total power received in the small time interval. For the MMC algorithm, $P_k = \sum_{i=1}^{N_k} p_{i,k}$. In the case of the CDMMC method, the contribution of first reflections to the total impulse response is calculated deterministically, similar to Barry's algorithm. Hence, there is no variance in that contribution. Variance exists for the contributions of second and remaining reflections only. Thus, (3.37) can be used to calculate relative estimated error in a small time interval for the CDMMC algorithm too, except P_k in this case is given by $P_k = \{\text{contribution of first reflection in } k\text{th time interval}\} + \sum_{i=1}^{N_k} p_{i,k}$.

From the above definitions, some observations can be made immediately. The CDMMC algorithm will show zero relative errors at time intervals when only the first reflections contribute to the total impulse response. Relative errors will be non-zero only when the second and subsequent reflections contribute to the total impulse response. This is a benefit of the CDMMC method compared to the MMC algorithm, where relative errors exist at all time intervals when power is received from any reflections. This benefit will be more useful when the first reflections contribute a large percentage of power to the total impulse response. For example, [9] shows that configuration C in Table 3.2 has an impulse response where the first reflections contribute 79.9% of the total power. Hence, zero relative errors for that portion of the total impulse response are more beneficial in terms of accuracy compared to other configurations such as configuration B in Table 3.2, where the first reflections contribute only 4.8% of the total power [9]. Thus, the CDMMC algorithm provides greater advantage when there is no LOS between the source and the receiver, so that first reflections contribute the most significant portion of power to the total impulse response.

It is also useful to compare relative estimated errors in the impulse responses calculated by both the MMC and the CDMMC algorithms when only the second and remaining reflections contribute to the total impulse response. The comparison should be done while keeping the total number of rays generated by each algorithm same, as this number can be used to compute the complexity of the algorithms later.

It is possible to make some initial guesses on the comparison of relative estimated errors based on the structure of the algorithms before showing simulation results. If the total number of rays generated by both the algorithms is the same, the CDMMC method should show higher relative estimated errors than the MMC method when there is a single source. The reason can be explained by noticing that all small surface elements in the CDMMC method are considered with equal importance in terms of number of rays generated from them. Since the main source is Lambertian, some surface elements will receive much less power as LOS contribution than others from the source, while some elements will receive much more, but they all generate the same number of rays. It means the values of the power contributions $p_{i,k}$ contributed by the N_k rays in the kth time interval in (3.37) vary more from each other in the CDMMC method than in the MMC algorithm. Hence, in the kth time interval, the term $\sum_{i=1}^{N_k} p_{i,k}^2$ will have higher values in the CDMMC method while the term $\sum_{i=1}^{N_k} p_{i,k}$ is approximately the same for both the algorithms since the latter term indicates total received power in the kth time interval. From (3.37) it can then be concluded that the CDMMC method should result in higher relative estimated errors for contributions of second and remaining reflections to the total impulse response than the MMC method if the total number of rays generated probabilistically is the same for both algorithms.

The values of relative estimated errors can be decreased by generating more rays from each small surface element in the CDMMC method, that is, to achieve the same relative estimated errors as the MMC algorithm, the CDMMC method will require more rays to be generated. Another approach is to assign the number of rays to be generated from the surface elements according to the power they receive as LOS contribution from the source instead of keeping the number same for all elements; however, this will increase the complexity of the method. The first approach, that is, increasing the number of rays, is automatically taken care of in the CDMMC algorithm when there are multiple sources. For example, if there are L sources in the room instead of one, the small surface elements can be assigned L power levels received from each source and L elapsed time periods that are required for light to traverse the distance between each of the sources and the surface elements. When a ray is generated from a surface element to calculate contributions of second and remaining reflections to the total impulse response, the power contributions and the elapsed times for L sources can be calculated easily by using these L values assigned to that surface element instead of generating L new rays. This means that the number of rays that actually contribute to the receiver is increased L times while the total number of rays generated probabilistically remains the same as the number of rays generated in the case of a single source. In the MMC algorithm, the number of rays generated from each source is decreased L times than the number used for a single source so that the total number of rays remains the same when either a single source or L sources are used. The end result is that values of relative estimated errors should be lower in the CDMMC method in the case of multiple sources compared to a single source and can even be lower than values of relative estimated errors achieved by the MMC method when the total number of rays generated by both methods remains same. In the latter case, the total number of rays generated by the MMC method can be increased to achieve the same values of relative estimated errors as the CDMMC algorithm.

As an illustration of the above discussions, Figure 3.17 shows relative estimated errors calculated for both the MMC and the CDMMC methods. The results are generated considering 10 reflections for each algorithm. For the CDMMC method, surface elements have an area of $1 \times 10^{-4}\,\text{m}^2$. Each element emits 10 rays. Figure 3.17a has been generated for configuration A of Table 3.2. For this configuration, the total number of rays is 17 350 000 for both the algorithms. Relative estimated errors for the CDMMC method are initially zero when only the first reflections contribute to the total impulse response. From Figure 3.16b, we can see that second and remaining reflections begin to contribute to the total impulse response at about 25 ns. Thus, from this starting point of time, relative errors of the CDMMC method exist and soon become greater than that of the MMC method, the reasons of which have been explained earlier. Figure 3.17b has been generated for a multiple source environment that has exactly the same configuration as configuration A of Table 3.2, except that four sources have been employed, whose locations are (2.5 m, 1.85 m, 3.3 m), (2.5 m, 3.7 m, 3.3 m), (5.0 m, 1.85 m, 3.3 m), and (5.0 m, 3.7 m, 3.3 m) given as (x, y, z) coordinates. All other parameters of the sources are the same as the single source in configuration A. As expected and explained earlier, from Figure 3.17b it can be seen that relative estimated errors of the CDMMC method in this case are less than that of the MMC method.

3.7.4.6 Computational Complexity

We compare the computational complexity of the algorithms in terms of elementary calculations. An elementary calculation can be defined as the calculation of received power by a surface element from a source using (3.6) and associated delay indicating the time required for light to traverse the distance between the source and the surface element. The number of elementary computations performed by Barry's algorithm [9] is approximately equal to N_e^K, where N_e is the number of surface elements and K is the number of reflections considered in the simulation. This value is very high and becomes quickly infeasible to compute impulse responses containing more than three reflections. The upper bound of elementary computations [24] performed by the MMC method is $N_{\text{comp}}^{\text{MMC}} = K N_{\text{rays\_MMC}} N_S$, where K is the number of reflections considered in the simulation, $N_{\text{rays\_MMC}}$ is the number of rays generated by the source, and N_S is the number of surfaces in the room. The upper bound of elementary computations performed by the CDMMC method is $N_{\text{comp}}^{\text{CDMMC}} = N_e + (K-1) N_e N_{\text{rays\_e}} N_S$, where N_e is the number of surface elements, K is the number of reflections considered in the simulation,

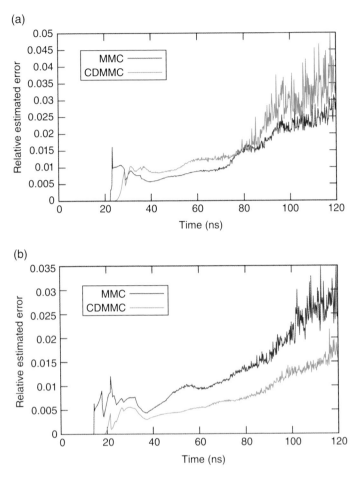

Figure 3.17 (a) Relative estimated errors of the MMC and the CDMMC methods generated for configuration A (Table 3.2) and (b) relative estimated errors of the MMC and the CDMMC methods generated for configuration A except with four sources whose locations are (2.5 m, 1.85 m, 3.3 m), (2.5 m, 3.7 m, 3.3 m), (5.0 m, 1.85 m, 3.3 m), and (5.0 m, 3.7 m, 3.3 m) given as (x, y, z) coordinates, and all other parameters of the sources are same as the single source in configuration A. Ten reflections are considered for both parts (a) and (b).

N_{rays_e} is the number of rays generated by each surface element, and N_s is the number of surfaces in the room. The first term of N_{comp}^{CDMMC} is the number of elementary computations performed for calculating the contribution of first reflections to the total impulse response. The number of elementary computations required to calculate contributions of the remaining reflections, that is, $K-1$ reflections to the total impulse response, is given by the second term of N_{comp}^{CDMMC}, which is similar to the form of N_{comp}^{MMC} as the value $N_e N_{rays\_e}$ indicates the total number of rays generated probabilistically by the CDMMC algorithm, whereas in the MMC method, this value is equal to N_{rays_MMC}. Note that an optimization can be performed in case of the CDMMC method when the source is Lambertian. Since a Lambertian source does not emit any light toward the opposite plane to which it is facing, all surface elements N_e need not be considered in the second term of $N_{\text{comp}}^{\text{CDMMC}}$. Surface elements belonging to the wall or the surface opposite to the plane to which the source is facing can be omitted as they will receive no power from the source as LOS contribution. However, there are sources that are non-Lambertian and light fixtures exist that have radiation patterns where a portion of the radiation is emitted from the opposite plane to which the source faces. Hence, in general, to calculate the upper bound, we consider all surface elements N_e in the second term of $N_{\text{comp}}^{\text{CDMMC}}$.

As explained in Section 3.7.4.5, the total number of rays needed by the MMC and the CDMMC algorithms to achieve the same relative estimated errors varies according to configurations. Hence, to compare the CDMMC method with the MMC algorithm, we first set $N_{\text{rays\_MMC}} = N_e N_{\text{rays\_e}}$, and then multiply the total number of rays required by the CDMMC method to achieve the same relative estimated errors as the MMC method by a parameter α. This parameter can take positive values either greater than or smaller than 1, depending on whether the total number of rays required by the CDMMC method is smaller or greater than the MMC method, respectively, to achieve the same relative estimated errors. If we now calculate the percentage increase or decrease of the number of elementary computations by the MMC method compared to the CDMMC algorithm, we have

$$
\begin{aligned}
&\frac{N_{\text{comp}}^{\text{MMC}} - N_{\text{comp}}^{\text{CDMMC}}}{N_{\text{comp}}^{\text{MMC}}} \times 100\% \\[2mm]
&= \frac{KN_{\text{rays\_MMC}}N_S - \{N_e + (K-1)\alpha N_e N_{\text{rays\_e}}N_S\}}{KN_{\text{rays\_MMC}}N_S} \times 100\% \\[2mm]
&= 1 - \alpha + \frac{\alpha}{K} - \frac{N_e}{KN_{\text{rays\_MMC}}N_S} \times 100\% \\[2mm]
&= 1 - \alpha\left(\frac{K-1}{K}\right) - \frac{N_e}{KN_{\text{rays\_MMC}}N_S} \times 100\%
\end{aligned}
\tag{3.38}
$$

We can neglect the last term in the final expression of (3.38) and focus on the first two terms. The value of $(K-1)/K$ is always less than unity. Hence, the value of $\alpha\left((K-1)/K\right)$ is less than the value of α. Thus, depending on the values of K and α, the final result can either be positive or negative. Hence, even if α is greater than 1, denoting that the CDMMC method requires more rays than the MMC method to achieve the same relative estimated errors, having lower K can yield positive results from (3.38), denoting that the MMC method will require more elementary computations than the CDMMC algorithm. A quick calculation shows that with the smallest number of reflections in the total impulse response that may be acceptable for lower data rate communication systems, that is, $K=3$, α can be as high as 1.5, that is, the CDMMC method can require 50% more rays than the MMC method but still can have lower computational complexity if only three reflections are considered. Higher values of α or higher number of reflections, K, will yield a negative result from (3.38), denoting that the MMC method will require less elementary computations than the CDMMC method. If α is smaller than 1, the result is always positive. As a numerical example, we take configuration B from Table 3.2 and plug in values into (3.38), while assuming $\alpha=1$. If the surface elements have an area of $1\times10^{-4}\,\text{m}^2$, we have $N_e = 1\ 735\ 000$. Other values are as follows: $K=5$, $N_S = 6$, $N_{\text{rays\_e}} = 10$. Putting these values in (3.38) yields the result of 19.67% decrease in elementary computations.

Thus, the MMC and the CDMMC methods are not directly comparable with each other in terms of elementary computations required by them as this is dependent on the configuration of the environment. The CDMMC method provides greater advantages when there is no LOS between the source and the receiver and when there are multiple sources. In other configurations that have a single source and LOS link between the source and the receiver, the CDMMC method may still be advantageous in terms of elementary computations if α is not too high and the number of reflections considered in the total impulse response is low.

3.7.5   Other Approaches for Impulse Response Calculation

Apart from the two main branches of algorithms for calculating impulse responses, that is, Barry's deterministic method and Monte Carlo ray-tracing, some other algorithms exist. A statistical approach is taken in Ref. [26] where the RMS delay spread and average delay are estimated from room parameters, and based on them, Rayleigh or Gamma distributions are used to fit the shape of the impulse response. However, it is not very accurate, and significant differences between the estimated and the real impulse responses are observed. An interesting method is presented in Ref. [27] where a geometric approach is taken. Instead of dividing the surfaces of the room into small reflecting elements, the loci of reflection points on a surface are found, which are used to calculate reflection points for the next order of reflections. The computing time required for this method is on the same order as required by Barry's method. Another approach reported in Ref. [28] directly estimates the frequency response of the channel from room parameters only instead of calculating the impulse response first. The frequency response that can be found by this approach is fairly acceptable in terms of accuracy for lower frequency regions, but at high frequency regions, the accuracy deteriorates. It should be noted that in all these methods, the light source is assumed to be monochromatic, that is, having a single wavelength. This allows for considering a fixed reflecting coefficient of the surfaces in the room. Monochromatic light can be generated from lasers; however, if LEDs are to be considered, the source becomes wideband in terms of power spectral density, and variation of the reflecting coefficients due to the wider range of wavelengths should be considered. This is the approach taken in Ref. [29] where LEDs are considered for VLC. The process used here is Barry's algorithm but the reflected power after hitting a surface is calculated by an integration of the spectral reflectance over the range of wavelengths emitted by the sources instead of just multiplying by a fixed reflecting coefficient as is done for monochromatic sources.

3.8   Conclusions

In this chapter, we have discussed why impulse response is an important tool in analyzing an indoor OWC as the channel impulse response can give us an estimate of ISI in case of high-speed communications through simulations. As calculating an impulse response for an indoor OWC is difficult because of reflections and multipath phenomena from different surfaces with different reflectivities, a number of algorithms exist to simulate with very good accuracy an impulse response of an indoor environment when all specifications of that system are available. Mainly, two branches of algorithms exist; deterministic approaches such as Barry's algorithm and probabilistic approaches based on Monte Carlo simulation such as the MMC method each has their advantages and disadvantages, including speed and accuracy. In this chapter, we have described these two main branches of algorithms, as well as a combined approach termed as the CDMMC algorithm that employs both deterministic and Monte Carlo methods. This approach is much faster than Barry's algorithm because the CDMMC method employs Monte Carlo simulations. Also it shows some advantages compared to the MMC method such as zero variance in the contribution of first reflections to the total impulse response and thus can provide greater advantages when no LOS exists between the source and the receiver, since in this case first reflections will contribute the most significant portion of power to the total impulse response. When there are multiple sources, the relative estimated errors of the CDMMC algorithm become less than that achieved by the MMC method due to multiple uses of each ray. The CDMMC method will require higher number of rays than the MMC method to achieve the same relative estimated errors in case of single source configurations; however, if this increase is not too significant, for example, less than 50% increase, and the number of reflections considered in the total impulse response is low, the CDMMC method may still be advantageous as the number of elementary computations becomes less than what is required by the MMC algorithm in such particular cases. Since the variance of both these algorithms is inversely proportional to the square root of the total number of rays as a property of the Monte Carlo method [21], it is possible to make a single run of these algorithms for a particular configuration and decide which algorithm will achieve the required relative estimated errors with lower number of elementary computations using (3.38) and hence achieve lower execution times.

References

[1] J. R. Barry, *Wireless Infrared Communications*. Springer US, Boston, MA, 1994.

[2] H. Hashemi, G. Yun, M. Kavehrad, F. Behbahani, and P. A. Galko, "Indoor propagation measurements at infrared frequencies for wireless local area networks applications," *IEEE Transactions on Vehicular Technology*, vol. 43, no. 3, pp. 562–576, 1994.

[3] J. M. Kahn, W. J. Krause, and J. B. Carruthers, "Experimental characterization of non-directed indoor infrared channels," *IEEE Transactions on Communications*, vol. 43, no. 234, pp. 1613–1623, 1995.

[4] J. Fadlullah and M. Kavehrad, "Indoor high-bandwidth optical wireless links for sensor networks," *Journal of Lightwave Technology*, vol. 28, no. 21, pp. 3086–3094, 2010.

[5] F. R. Gfeller and U. Bapst, "Wireless in-house data communication via diffuse infrared radiation," *Proceedings of the IEEE*, vol. 67, no. 11, pp. 1474–1486, 1979.

[6] P. Beckmann and A. Spizzichino, *The Scattering of Electromagnetic Waves from Rough Surfaces*. Pergamon Press, Gosford, NY, 1963.

[7] B. T. Phong, "Illumination for computer generated pictures," *Communications of the ACM*, vol. 18, no. 6, pp. 311–317, 1975.

[8] C. R. Lomba, R. T. Valadas, and A. d. O. Duarte, "Experimental characterisation and modelling of the reflection of infrared signals on indoor surfaces," *IEE Proceedings—Optoelectronics*, vol. 145, no. 3, pp. 191–197, 1998.

[9] J. R. Barry, J. M. Kahn, W. J. Krause, E. A. Lee, and D. G. Messerschmitt, "Simulation of multipath impulse response for indoor wireless optical channels," *IEEE Journal on Selected Areas in Communications*, vol. 11, no. 3, pp. 367–379, 1993.

[10] Z. Zhou, C. Chen, and M. Kavehrad, "Impact analyses of high-order light reflections on indoor optical wireless channel model and calibration," *Journal of Lightwave Technology*, vol. 32, no. 10, pp. 2003–2011, 2014.

[11] F. J. López-Hernández and M. J. Betancor, "DUSTIN: algorithm for calculation of impulse response on IR wireless indoor channels," *Electronics Letters*, vol. 33, no. 21, pp. 1804–1806, 1997.

[12] J. B. Carruthers and P. Kannan, "Iterative site-based modeling for wireless infrared channels," *IEEE Transactions on Antennas and Propagation*, vol. 50, no. 5, pp. 759–765, 2002.

[13] J. B. Carruthers, S. M. Carroll, and P. Kannan, "Propagation modelling for indoor optical wireless communications using fast multi-receiver channel estimation," *IEE Proceedings—Optoelectronics*, vol. 150, no. 5, pp. 473–481, 2003.

[14] M. Abtahi and H. Hashemi, "Simulation of indoor propagation channel at infrared frequencies in furnished office environments," in Personal, Indoor and Mobile Radio Communications, 1995. PIMRC'95. Sixth IEEE International Symposium on Wireless: Merging onto the Information Superhighway, pp. 306–310, vol. 1, 1995, Toronto, Canada, September 27–29, 1995.

[15] F. Li, X. Li, W. Zou, and J. Chen, "Simplified calculation method of indoor optical impulse response based on recursive algorithm," in 2012 1st IEEE International Conference on Communications in China Workshops (ICCC), pp. 49–53, 2012, Beijing, China, August 15–17, 2012.

[16] Y. A. Alqudah and M. Kavehrad, "MIMO characterization of indoor wireless optical link using a diffuse-transmission configuration," *IEEE Transactions on Communications*, vol. 51, no. 9, pp. 1554–1560, 2003.

[17] F. J. López-Hernández, R. Pérez-Jiménez, and A. Santamaría, "Monte Carlo calculation of impulse response on diffuse IR wireless indoor channels," *Electronics Letters*, vol. 34, no. 12, pp. 1260–1262, 1998.

[18] F. J. López-Hernández, R. Pérez-Jiménez, and A. Santamaría, "Modified Monte Carlo scheme for high-efficiency simulation of the impulse response on diffuse IR wireless indoor channels," *Electronics Letters*, vol. 34, no. 19, pp. 1819–1820, 1998.

[19] F. J. López-Hernández, R. Pérez-Jiménez, and A. Santamaría, "Novel ray-tracing approach for fast calculation of the impulse response on diffuse IR-wireless indoor channels," *Proceedings—SPIE Optical Wireless Communications II*, vol. 3850, no. 1, pp. 100–107, 1999.

[20] F. J. López-Hernández, R. Pérez-Jiménez, and A. Santamaría, "Ray-tracing algorithms for fast calculation of the channel impulse response on diffuse IR wireless indoor channels," *Optical Engineering*, vol. 39, no. 10, pp. 2775–2780, 2000.

[21] O. González, S. Rodríguez, R. Pérez-Jiménez, B. R. Mendoza, and A. Ayala, "Error analysis of the simulated impulse response on indoor wireless optical channels using a Monte Carlo-based ray-tracing algorithm," *IEEE Transactions on Communications*, vol. 53, no. 1, pp. 124–130, 2005.

[22] S. P. Rodríguez, R. P. Jiménez, B. R. Mendoza, F. J. L. Hernández, and A. J. A. Alfonso, "Simulation of impulse response for indoor visible light communications using 3D CAD models," *EURASIP Journal on Wireless Communications and Networking*, vol. 2013, no. 1, pp. 1–10, 2013.

[23] M. Zhang, Y. Zhang, X. Yuan, and J. Zhang, "Mathematic models for a ray tracing method and its applications in wireless optical communications," *Optics Express*, vol. 18, no. 17, pp. 18431–18437, 2010.

[24] O. González, S. Rodríguez, R. Pérez-Jiménez, B. R. Mendoza, and A. Ayala, "Comparison of Monte Carlo ray-tracing and photon-tracing methods for calculation of the impulse response on indoor wireless optical channels," *Optics Express*, vol. 19, no. 3, pp. 1997–2005, 2011.

[25] H.-S. Lee, "A photon modeling method for the characterization of indoor optical wireless communication," *Progress in Electromagnetics Research*, vol. 92, pp. 121–136, 2009.

[26] R. Pérez-Jiménez, J. Berges, and M. J. Betancor, "Statistical model for the impulse response on infrared indoor diffuse channels," *Electronics Letters*, vol. 33, no. 15, pp. 1298–1300, 1997.

[27] D. Mavrakis and S. R. Saunders, "A novel modelling approach for wireless infrared links," in The Third International Symposium on Wireless Personal Multimedia Communications, pp. 609–614, 2000, Bangkok, Thailand, November 12–15, 2000.

[28] V. Jungnickel, V. Pohl, S. Nönnig, and C. von Helmolt, "A physical model of the wireless infrared communication channel," *IEEE Journal on Selected Areas in Communications*, vol. 20, no. 3, pp. 631–640, 2002.

[29] K. Lee, H. Park, and J. R. Barry, "Indoor channel characteristics for visible light communications," *IEEE Communications Letters*, vol. 15, no. 2, pp. 217–219, 2011.

4

Analyses of Indoor Optical Wireless Channels Based on Channel Impulse Responses

4.1 Introduction

In the last chapter, we have included details on techniques to simulate the channel impulse response of an indoor optical wireless channel (OWC). With the channel impulse response, it is possible to analyze a communications link and predict its performance in the presence of noise and other factors that influence bit-error-rate performance. In this chapter, we aim to first analyze in detail different indoor channel impulse responses and their Fourier transforms, and the frequency responses of the channels, and show that the size of the room and the positions of transmitters and receivers have a big influence on their achievable performance. Later, we will analyze the different types of noise present in indoor wireless communication systems. We will also show the effects of furniture and other reflecting surfaces in a room on the channel impulse response compared to a room having no furniture.

4.2 Analyses of Optical Wireless Channel Impulse Responses

Indoor OWC impulse responses are very much dependent on the configuration of the room and positions and configurations of transmitters and receivers. To show this, we need to evaluate impulse responses of many different rooms having transmitters and receivers at many different positions within them. We can broadly select three room configurations for our analyses, making the lengths, the widths, and the heights of the rooms representative of usual room configurations at normal dwellings, at office buildings, and for larger conference rooms, respectively. We will assume the walls of the rooms are made with plaster and colored white, so that the reflecting coefficient is 0.7 for all walls and the ceiling, and for the floor we will assume it is 0.25. We will confine our analyses here for monochromatic systems, that is, we will assume the transmitter is a laser source emitting only one wavelength. Analyses where the

Short-Range Optical Wireless: Theory and Applications, First Edition. Mohsen Kavehrad, M. I. Sakib Chowdhury and Zhou Zhou.

source is a light emitting diode (LED) follow a similar method, except that when the impulse response is calculated, different reflecting coefficients at different wavelengths have to be taken care of since LED is a broad linewidth source. Also, since LEDs have a limited bandwidth of around 20–100 MHz, it is more useful to analyze impulse responses of indoor OWCs where the source is laser as the latter type of sources have a much higher bandwidth, in the tens of gigahertz range. At the receiver side, the area and the field-of-view (FOV) of the photodiode are important parameters to consider. The FOV will control directly the amount of reflections that the photodiode receives; hence by modifying the FOV, it is possible to consider a communications system where inter-symbol-interference (ISI) is low. The FOV is controllable by the lens that is usually employed at the opening of the detector active area, that is, it is a modifiable parameter. What is not modifiable is the detector active area which controls the amount of optical power that the detector receives to convert to electrical current. High-speed photodiodes have a small detection area, and the higher the supported bandwidth of the photodiode is, the smaller the area of the detection surface becomes. It means for high-speed photodiodes, the intensity of light reaching the receiver has to be more to maintain the same performance as low-speed photodiodes. One possible way to ensure this is to use a focusing lens and concentrators in front of detectors that capture more light. We will first analyze the systems without a focusing lens or concentrator at the receiver side.

Table 4.1 shows the room configurations that will represent three categories of rooms, namely, those at home, at office, and at conference halls. At this point, we consider there is no furniture present in the rooms.

The conventions of the coordinate system to be followed for source and receiver locations are the same as shown in Figure 3.4. Since we will analyze impulse responses for different locations of sources and receivers, we will not mention their locations in a fixed tabular format at this point. The fixed parameters of the source and the receiver are given in Table 4.2.

The elevation of the source will indicate whether the communication link is a line-of-sight (LOS) or non-line-of-sight (NLOS) system. We will place the receiver at a height of 0.8 m, indicating that it is placed on some fixture such as a table. The elevation of the receiver is +90° which means it is facing upward toward the ceiling. It may be helpful to recall that the elevation of a source or a receiver is defined as the angle that the normal of the source or the receiver makes with the xy-plane. Now, since the receiver faces upward in the following analyses, the source's elevation and the z coordinate will indicate the type of link. If the z

Table 4.1 Room configurations for impulse response analyses

	Configuration A	Configuration B	Configuration C
Length	5 m	6 m	7.5 m
Width	4 m	5 m	5.5 m
Height	2.5 m	3.5 m	3.5 m
Reflection coefficients	ρ_{north}: 0.7	ρ_{north}: 0.7	ρ_{north}: 0.7
	ρ_{south}: 0.7	ρ_{south}: 0.7	ρ_{south}: 0.7
	ρ_{east}: 0.7	ρ_{east}: 0.7	ρ_{east}: 0.7
	ρ_{west}: 0.7	ρ_{west}: 0.7	ρ_{west}: 0.7
	$\rho_{ceiling}$: 0.7	$\rho_{ceiling}$: 0.7	$\rho_{ceiling}$: 0.7
	ρ_{floor}: 0.25	ρ_{floor}: 0.25	ρ_{floor}: 0.25

Table 4.2 Source and receiver configurations for impulse response analyses

Source	Receiver
Lambertian mode: 1	Area: $2.5 \times 10^{-9}\,m^2$
x: to be varied	x: to be varied
y: to be varied	y: to be varied
z: to be varied	z: 0.8 m
Elevation: −90° (LOS)/+90° (NLOS)	Elevation: +90°
Azimuth: 0°	Azimuth: 0°
	FOV: 70°

coordinate of the source is almost equal to the room height and its elevation is −90°, the source is facing downward from the ceiling toward the floor. This type of link can be an LOS link if the receiver's FOV and position enable it to receive light from the source directly. Since in this chapter our analyses are confined to monochromatic sources, it means the actual laser transmitter is located elsewhere, but the laser beam hits the ceiling at the position indicated. The location of the laser transmitter is of no concern to the system because the impact point where the emitted beam has hit the ceiling acts as the source for the communication system. Hence, whether the link is LOS or NLOS will be indicated by this impact point acting as a source. On the other hand, if the z coordinate of the transmitter is about the same as the receiver and elevation of the transmitter is +90°, it will indicate that it is facing upward toward the ceiling. This is possible for a monochromatic system if a source is made by placing a diffuser in front of a laser transmitter. The diffuser makes the output radiation pattern Lambertian, and the whole transmitter along with the diffuser can be made to face directly upward. Thus, it is essentially an NLOS system as both the source and the receiver face upward, and the receiver can detect light from reflections only. Thus, in this chapter, when we specify a link as non-directed LOS, it means the source is actually an impact point of a laser beam on the ceiling; when we specify a link as non-directed NLOS, it means the source is the actual laser transmitter that has a diffuser in front of it and is facing upward toward the ceiling. In both cases, the receiver will be facing upward too.

The area of the receiver is also an important parameter as described earlier. Here, we have assumed a realistic value to be the detection area for a photodiode having bandwidth of about 1 GHz and having a detectable wavelength range of 850–1600 nm (e.g., [1]). The selected value is quite small, but is necessary if we want to transmit at high data rates requiring high bandwidth photodiode. If the detectable wavelength range of the photodiode is 400–1000 nm, the detector area can be somewhat higher with approximately the same bandwidth (e.g., [2]). Detection area actually acts as a multiplier when the total impulse response is calculated, according to (3.6), and hence the impulse response for a different photodiode having a different detection area can be derived from a given impulse response simply by first dividing the magnitude of the impulse response by the given detection area and then multiplying it by the new detection area of the new photodiode.

We will first analyze some channel conditions where a non-directed LOS link exists between the source and the receiver. We will discuss root-mean-square (rms) delay spread—an important parameter relating impulse responses to multipath induced ISI. We will also discuss effects of FOV on rms delay spread. Later we will show these analyses for non-directed NLOS links also.

4.2.1 Non-Directed LOS Links

We will show impulse responses for two positions of the receiver assuming the source is on the ceiling directly in the middle of the room. The two positions will be in the middle of the room, that is, directly underneath the source and at the corner of the room. These two positions will give us some idea on the nature of impulse responses in a typical indoor room environment where non-directed LOS links exist between the source and the receiver.

Figure 4.1 shows the impulse response in the room of configuration A of Table 4.1 when the receiver coordinate is (2.5 m, 2.0 m, 0.8 m) and the source coordinate is (2.5 m, 2.0 m, 2.5 m). This denotes that the receiver is directly underneath the source and both of them are exactly in the middle of the room. It is helpful to recall that impulse responses in optical wireless systems have the unit of power (W). The LOS response in this case as shown in the figure is 2.7535×10^{-10} W. Also it is helpful to recall that the source power is assumed to be 1 W when the impulse response has been calculated. Thus, because of the small detector area and the square-law dependence of path loss, the received power has become quite low. The rest of the impulse response is calculated for reflected lights only, and we can see that compared to the LOS response, the reflected response is much smaller. The impulse response has been calculated for 10 reflections.

Figure 4.2 shows the impulse response in the room of configuration A of Table 4.1 when the receiver coordinate is (0.1 m, 0.1 m, 0.8 m) and the source coordinate is (2.5 m, 2.0 m, 2.5 m). This denotes that the receiver is at one of the corners of the room while the source is in the middle of room. In this case, the LOS response becomes much smaller while responses from reflections become greater in magnitude than the impulse response shown in Figure 4.1. The reason for this is that the distance between the source and the receiver has increased for the configuration of Figure 4.2 compared to the configuration of Figure 4.1. Also, since the

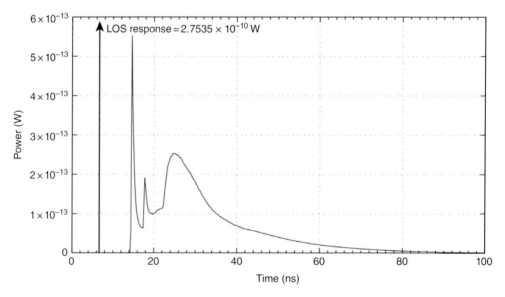

Figure 4.1 Impulse response of a non-directed LOS link of configuration A of Table 4.1, when source coordinate is (2.5 m, 2.0 m, 2.5 m) and receiver coordinate is (2.5 m, 2.0 m, 0.8 m).

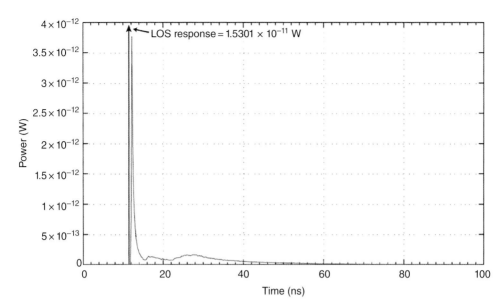

Figure 4.2 Impulse response of a non-directed LOS link of configuration A of Table 4.1, when source coordinate is (2.5 m, 2.0 m, 2.5 m) and receiver coordinate is (0.1 m, 0.1 m, 0.8 m).

receiver is much nearer to the walls, the reflections are able to contribute a higher amount of power to the receiver. Hence, from these two figures, we can draw a conclusion that from the shape of an impulse response, we can roughly estimate the relative positions of the source and the receiver within a room for a non-directed LOS system.

Let us analyze the same two relative positions of the source and the receiver for configuration B of Table 4.1. Configuration B is a larger room than configuration A. So, it is expected that we should receive more reflected responses having higher time delays than that of configuration A. We see evidence of this in Figure 4.3, where the receiver is placed directly underneath the source and both the source and the receiver are exactly in the middle of the room. The "tail" of the impulse response goes well beyond 100 ns which is higher than the delays observed for configuration A in Table 4.1. We also see that the value of the LOS response in Figure 4.3 is smaller than configuration A shown in Figure 4.1 as the distance between the source and the receiver is greater in configuration B than that of configuration A.

Figure 4.4 shows the impulse response when the receiver is placed at the corner of the room of configuration B, keeping the source at the same position as in Figure 4.3. The differences between these two impulse responses are similar in nature as we observed in Figures 4.1 and 4.2. The LOS response becomes a bit smaller than what is observed in Figure 4.3 since the distance between the source and the receiver is larger. The reflections contribute comparatively more to the total impulse response, and the delays are shorter too, since the receiver is placed much closer to the walls from where the reflections are actually received. Comparing Figures 4.2 and 4.4, we still observe that the "tail" of the impulse response is longer for configuration B than configuration A because of the larger room size.

If we continue to calculate impulse responses for configuration C of Table 4.1, we should observe similar features and differences. Figure 4.5 shows the impulse response calculated for

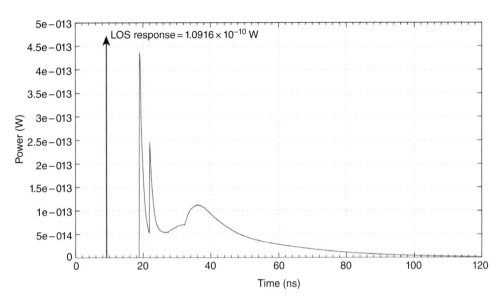

Figure 4.3 Impulse response of a non-directed LOS link of configuration B of Table 4.1, when source coordinate is (3.0 m, 2.5 m, 3.5 m) and receiver coordinate is (3.0 m, 2.5 m, 0.8 m).

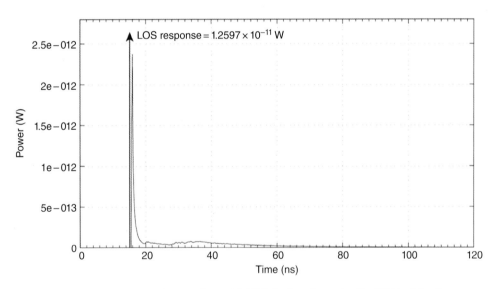

Figure 4.4 Impulse response of a non-directed LOS link of configuration B of Table 4.1, when source coordinate is (3.0 m, 2.5 m, 3.5 m) and receiver coordinate is (0.1 m, 0.1 m, 0.8 m).

configuration C of Table 4.1, which consists of the largest room size among the three configurations, where the source and the receiver are placed in the middle of the room. We can immediately observe that since the room height of configurations B and C are the same, the LOS response is also exactly the same (from Figure 4.3) because of the same distance between the source and the receiver. Though not immediately distinguishable from observing

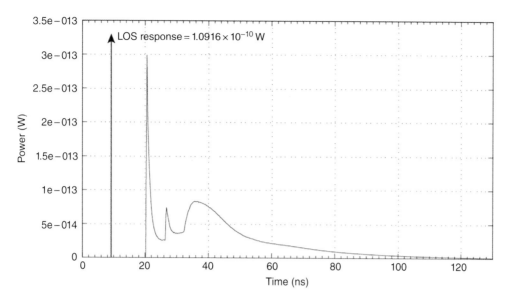

Figure 4.5 Impulse response of a non-directed LOS link of configuration C of Table 4.1, when source coordinate is (3.75 m, 2.75 m, 3.5 m) and receiver coordinate is (3.75 m, 2.75 m, 0.8 m).

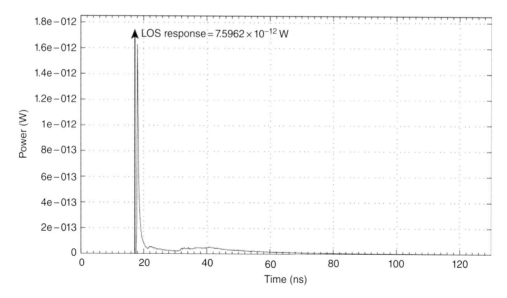

Figure 4.6 Impulse response of a non-directed LOS link of configuration C of Table 4.1, when source coordinate is (3.75 m, 2.75 m, 3.5 m) and receiver coordinate is (0.1 m, 0.1 m, 0.8 m).

Figure 4.5, the "tail" of the impulse response is also actually longer than that of configuration B because of the larger room size.

Similar to the other configurations, we can place the receiver at the corner of the room of configuration C and calculate the impulse response, which is shown in Figure 4.6. The LOS response becomes progressively smaller than other configurations as the distance between

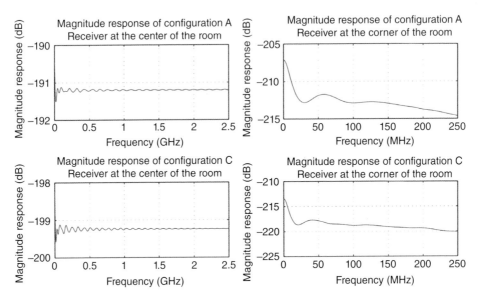

Figure 4.7 Magnitude responses calculated from impulse responses of LOS channels simulated for two locations of the receiver for configurations A and C of Table 4.1.

the source and the receiver increases. Also, the reflections are much closer to the LOS response as the receiver has been placed near the walls. Hence, from these impulse responses, all of which have been calculated for a non-directed LOS link, that is, the receiver has a LOS link with the source, we can draw some conclusions. First, the magnitude of the LOS impulse response is usually much higher than the contributions of the reflections to the total impulse response. This is true when the source is in the middle of the room and the receiver is directly underneath the source. As the receiver is moved toward the corners of the room, the LOS response becomes smaller and the contributions of the reflections to the total impulse response become larger, but they also become much closer in time domain. Hence, for all cases, in a non-directed LOS channel, it may be possible to neglect the reflections when the performance of an optical wireless communications system is considered, but only for low data rates. For high data rate systems, the small values of contributions of the reflections to the total impulse response compared to the LOS response may still influence the performance significantly by introducing ISI.

We will now show the magnitude responses calculated from the impulse responses simulated for two locations of the receiver for configurations A and C of Table 4.1. From the magnitude responses, we can find the −3 dB bandwidth of the channels at those locations, which will be useful in our present discussion. The magnitude responses are shown in Figure 4.7.

The −3 dB bandwidths observed from the magnitude responses of Figure 4.7 are given in Table 4.3.

When the receiver is at the center of the room, the −3 dB bandwidth is high, which is to be expected from the smallest distance between the source and the receiver and lowest contributions of reflections to the total impulse response at these locations. When the receiver is at the corner, we observe decreasing −3 dB bandwidth as the room size increases, that is, the distance

Table 4.3 $-3\,dB$ bandwidth of LOS channels consisting of two locations of the receiver for configurations A and C of Table 4.1

	Receiver at the center of the room	
	Configuration A	Configuration C
$-3\,dB$ Bandwidth	>2.5 GHz	>2.5 GHz
	Receiver at the corner of the room	
	Configuration A	Configuration C
$-3\,dB$ Bandwidth	11.73 MHz	8.55 MHz

between the source and the receiver increases. Since this distance in an LOS link defines the magnitude of the DC component in the magnitude response, it is expected that as this distance in an LOS link increases and at the same time as the reflections contribute comparatively more power to the total impulse response, the $-3\,dB$ bandwidth also decreases.

At this point, it will be helpful to introduce an important parameter that quantifies the severity of ISI introduced by a multipath channel—rms delay spread. The rms delay spread can be computed from the impulse response $h(t)$ by

$$D = \sqrt{\frac{\int_{-\infty}^{\infty}(t-\mu)^2 h^2(t)dt}{\int_{-\infty}^{\infty}h^2(t)dt}} \qquad (4.1)$$

Here μ is the mean delay which can be calculated by

$$\mu = \frac{\int_{-\infty}^{\infty} t\, h^2(t)dt}{\int_{-\infty}^{\infty}h^2(t)dt} \qquad (4.2)$$

As we have discussed earlier, the impulse response of an indoor OWC $h(t)$, and hence the delay spread D computed from it, can be considered to be deterministic quantities. This is because as long as the source, the receiver, and other reflectors within the environment do not change their positions, the impulse response and the delay spread will not change. The rms delay spread is deeply related to quantifying the severity of ISI in the sense that the inverse of the delay spread is proportional to the coherence bandwidth of the channel [3, 4], that is, the bandwidth over which the channel can be considered "flat." Within the coherence bandwidth of the channel, the frequencies experience a similar amount of attenuation, that is, there is absence of frequency selectivity, and hence, there should be no inter-symbol interference (ISI) in time domain. It is accepted that the maximum bit-rate that can be transmitted through a channel without introducing ISI and without using an equalizer at the receiver end can be approximately related to the rms delay spread as [4], $R_b \leq 1/10D$.

In Table 4.4, we show some quantities related to the impulse responses calculated for configurations A and C of Table 4.1 for non-directed LOS links. We show the LOS power

Table 4.4 Power contributions from LOS responses and reflections, percentage of total power contributed by reflections, and rms delay spread calculated from impulse responses of non-directed LOS links of configurations A and C of Table 4.1 at two receiver positions when the source is at the center of the ceiling

		Configuration A	Configuration C
Power contribution from LOS response	Receiver at the center	2.7535×10^{-10} W	1.0916×10^{-10} W
	Receiver at the corner	1.5301×10^{-11} W	7.5962×10^{-12} W
Power contribution from reflections	Receiver at the center	2.5009×10^{-11} W	1.2961×10^{-11} W
	Receiver at the corner	2.889×10^{-11} W	1.3752×10^{-11} W
Percentage of total power contributed by reflections	Receiver at the center	8.3263%	10.6131%
	Receiver at the corner	65.376%	64.4173%
Root-mean-square delay spread	Receiver at the center	0.1557 ns	0.2292 ns
	Receiver at the corner	1.4668 ns	1.6051 ns

contribution, the contribution of reflections, and the percentage of total power contributed by the reflections, as well as rms delay spread.

From Table 4.4, we can see that when the receiver is in the middle of the room directly underneath the source, the LOS contribution is typically much higher (at least one order of magnitude) than contributions from reflections. The percentage of total power contributed by reflections is also around 8–10%. Since reflections do not contribute much to the total impulse response, the rms delay spread remains small, typically smaller than approximately 0.3 ns. On the other hand, when the receiver has been placed at the corner of the room, we see very different behavior in terms of power contributions by reflections. The LOS response contribution and the contribution of reflections become almost equal in these cases, and to be specific, actually the LOS response becomes smaller than contributions of reflections in all cases. This is evident from the percentage of total power contributed by reflections where the percentage value is always greater than approximately 60%. Correspondingly, the rms delay spread is much worse, about approximately 1.4–1.6 ns. From these delay spread values, it can be concluded that these channels will support a bit rate of approximately 350–667 Mbps without introducing ISI and without equalization at the receiver end when the receiver is placed in the middle of the room directly underneath the source. When the receiver is at the corner, the supported bit rate, without introducing ISI and without requiring equalizer at the receiver end becomes 60–70 Mbps.

We can show the effects of increasing rms delay spread more clearly as the receiver moves toward the corner. Figure 4.8 shows rms delay spreads at every receiver location inside the room of configuration B of Table 4.1 when the source is at the center of the ceiling. We can observe minimum rms delay spread when the distance between the receiver and the source is the smallest, in this case, when the receiver is directly underneath the source. As the distance increases, and as the receiver moves closer to reflecting surfaces, that is, walls of the room, rms delay spread increases. Hence, from all these discussions, it is evident that the highest data rate without ISI that can be transmitted in an optical wireless system without employing equalizers at the receiver end is dependent on factors such as proximity of the receiver to reflecting surfaces and the distance between the receiver and the source. The smallest rms delay spread, or in other words, the highest data rate without ISI and without equalizers at the

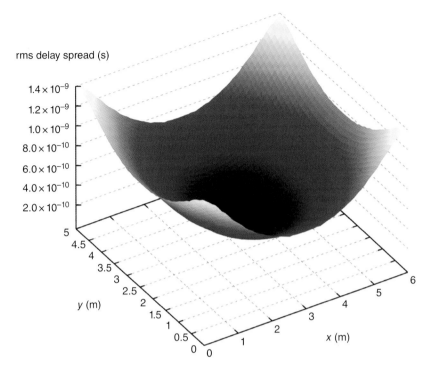

rms delay spread (s)

Figure 4.8   rms delay spreads calculated for non-directed LOS links at every receiver location at an *xy*-plane of height 0.8 m inside the room of configuration B of Table 4.1 when the source is at the center of the ceiling.

receiver end can be achieved when this distance is minimum in the system and when the receiver is farthest away from reflecting surfaces.

Let us consider some cases when the source is not placed in the middle of the ceiling. If the source is placed elsewhere, it is possible that the link may become a non-directed NLOS one instead of a non-directed LOS link, depending on the receiver location and receiver FOV. For example, the receiver that we have been using until now from Table 4.2 has a FOV of 70° and was placed at a height of 0.8 m. We can calculate the radius of the circle on the ceiling within which any source has an LOS link with the receiver. For configurations B and C of Table 4.1, the height of the room is 3.5 m. Hence, the radius will be $r = (3.5 - 0.8)\tan(70°)\,\text{m} = 7.42\,\text{m}$. Since the length of the largest room (configuration C) is only 7.5 m, we can safely assume that with this FOV of the receiver, wherever the source is placed at the ceiling in a typical operation of the communications system, there will always be an LOS link with the receiver. Of course, if both the source and the receiver are placed on opposite corners of the room, that is, if the horizontal distance between them becomes larger than the calculated radius above, there will be no LOS link. Figure 4.9 shows such a scenario where the source is out of the FOV of the receiver, and the link becomes a non-directed NLOS link.

Let us reduce the FOV of the receiver to 60° and recalculate the radius. The radius in this case becomes $r = (3.5 - 0.8)\tan(60°)\,\text{m} = 4.67\,\text{m}$. Since this value is not very high, we can assume that such a horizontal distance between the receiver and the source can often be

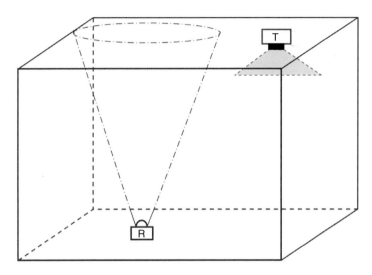

Figure 4.9 Special case showing a non-directed NLOS link because of small receiver FOV and large horizontal distance between the source and the receiver when the source is on the ceiling.

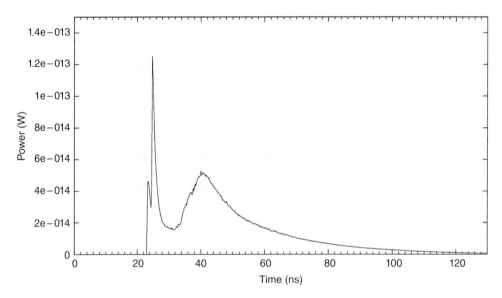

Figure 4.10 Impulse response of a non-directed NLOS link of configuration C of Table 4.1, when receiver FOV is 60°, source coordinate is (6.0 m, 2.75 m, 3.5 m), and receiver coordinate is (0.5 m, 0.5 m, 0.8 m).

exceeded. We have shown an impulse response of such a case in Figure 4.10, where the source is placed at coordinate (6.0 m, 2.75 m, 3.5 m) and the receiver coordinate is (0.5 m, 0.5 m, 0.8 m) inside the room of configuration C. We have reduced the FOV of the receiver to 60°. As we have calculated, this is a non-directed NLOS link, and hence we do not observe any LOS response. The total impulse response comprises contributions from reflections only, and thus the delay spread is expected to be high compared to non-directed LOS links. The delay spread for this impulse response is 12.6356 ns, which is much larger than the values shown in Table 4.4.

It is thus clear that FOV is also an important parameter in determining the rms delay spread of an optical wireless link. Larger FOVs will capture more light from emissions and hence the delay spread will be larger. Smaller FOVs will capture light from a smaller area, and correspondingly the system performance at high data rates will be better due to decrease in rms delay spread. However, smaller FOVs also can introduce a much larger rms delay spread if the source moves out of the receiver's viewing area. This is possible since a small FOV indicates a smaller horizontal distance between the source and the receiver that ensures an LOS link, and thus the requirement of moving either the source or the receiver so that an LOS link exists becomes important. We will show some effects of changing the FOV of the receiver for configuration C in Table 4.1 when the source coordinate is (6.0 m, 2.75 m, 3.5 m). The receiver will be placed at three different locations, (3.75 m, 2.75 m, 0.8 m), (1.5 m, 2.75 m, 0.8 m), and (0.5 m, 0.5 m, 0.8 m), where we will change the FOV from 10° to 85° in an increment of 5° at each location and plot the calculated rms delay spread from the simulated impulse responses.

Figure 4.11 shows the effects that receiver FOVs have on rms delay spread at the first receiver location, that is, (3.75 m, 2.75 m, 0.8 m). We can observe that from 10° to 35°, the delay spread is higher compared to other FOVs. Actually this signifies that at those FOVs, the link is non-directed NLOS instead of LOS. Also we can see that from 10° to 35°, the rms delay spread gradually increases. The reason for this can be explained by realizing that a larger FOV also means more contributions to the total impulse response by reflections since the receiver can capture more light. At 40° the link becomes non-directed LOS, and hence the rms delay

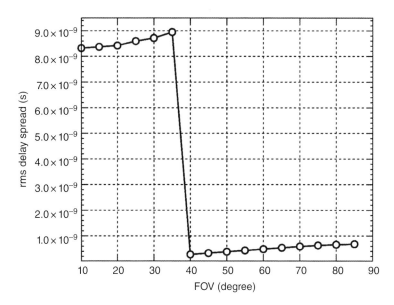

Figure 4.11 Effect of receiver FOV on rms delay spread of a non-directed LOS link when source coordinate is (6.0 m, 2.75 m, 3.5 m) and receiver coordinate is (3.75 m, 2.75 m, 0.8 m) in the room of configuration C of Table 4.1.

spread greatly reduces. Again, from 40° to 85°, we see a gradual increase in rms delay spread as the receiver captures more light from reflections.

Figure 4.12 shows the effects that receiver FOV causes on rms delay spread at the second receiver location, that is, (1.5 m, 2.75 m, 0.8 m). The difference between this location and the previous one is that the horizontal distance between the source and the receiver is larger in this case. Because of this, it is to be expected that larger values of FOV than the previous case will lead to non-directed NLOS links. This is precisely the behavior that we observe in Figure 4.12, where the link is non-directed NLOS for FOVs 10–55°. At 60° the link becomes a non-directed LOS link and a large decrease in rms delay spread is seen. Also the rms delay spread at every FOV is larger in Figure 4.12 than the configuration of Figure 4.11 since the receiver is farther away from the source, and reflections contribute more to the total impulse response. Gradual increase in rms delay spread within the ranges 10–55° and 60–85° is also observed, similar to the previous case, which carries the same explanation of the receiver capturing more light from reflections at higher FOVs.

Finally, we place the receiver near the corner at location (0.5 m, 0.5 m, 0.8 m) and show the effects that receiver FOV causes on rms delay spread, in Figure 4.13. The overall behavior is similar to that of the previous two cases. Following the same trend, the range of FOVs when the link is non-directed, NLOS has increased up to 65°, as at this location the receiver has the largest horizontal distance from the source among the three locations we are considering. The values of rms delay spreads are also larger compared to the other two cases because of the same reason. Also, gradual increase in rms delay spread as FOV increases within the two ranges of FOV can be seen from the figure.

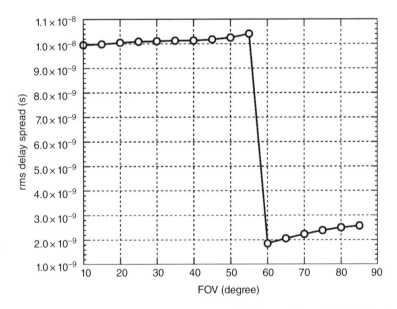

Figure 4.12   Effect of receiver FOV on rms delay spread of a non-directed LOS link when source coordinate is (6.0 m, 2.75 m, 3.5 m) and receiver coordinate is (1.5 m, 2.75 m, 0.8 m) in the room of configuration C of Table 4.1.

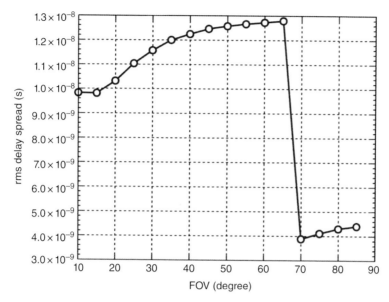

Figure 4.13 Effect of receiver FOV on rms delay spread of a non-directed LOS link when source coordinate is (6.0 m, 2.75 m, 3.5 m) and receiver coordinate is (0.5 m, 0.5 m, 0.8 m) in the room of configuration C of Table 4.1.

Hence, it is important to note that though a smaller FOV of the receiver will result in smaller rms delay spread if there exists an LOS response, the system may perform much worse when that LOS link is no longer present. We can compare rms delay spreads at FOV of 40° in Figures 4.11, 4.12, and 4.13 to observe this more prominently. Thus, it is necessary to reach a compromise according to the room size and possible locations of the source and the receiver and decide what receiver FOV may yield the best, that is, the smallest rms delay spreads, in most of the application scenarios in that room configuration.

Performance of an indoor optical wireless link does not only depend on rms delay spread, however. Signal-to-noise ratio (SNR) also plays a vital role in determining the performance metrics of a system. We will discuss noise sources later in this chapter. However, at this point, it may be helpful to define optical path loss. Path loss is measured as the ratio of the receiver power to the transmitted power, and hence gives an estimate of possible link budgets that can satisfy some performance criteria. As impulse responses of optical wireless systems are simulated assuming the transmitted power from the source to be 1 W, we can define path loss as

$$\text{Path loss}\,(\text{dB}) = -10\log_{10}\left(\int_0^\infty h(t)\,dt\right) \tag{4.3}$$

Having simulated the impulse responses of every location of configuration B of Table 4.1 at an xy-plane of height 0.8 m when the source is at the center of the ceiling, we can calculate path losses using (4.3) and show the results. Figure 4.14 shows these results. The figure is quite similar to Figure 4.8, as expected, because higher rms delay spread typically means

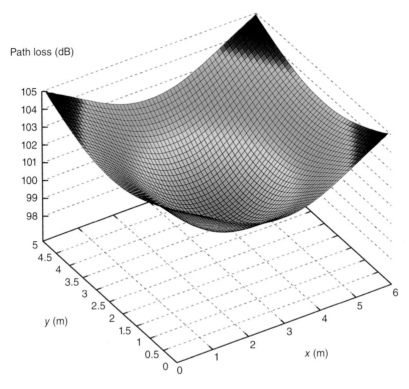

Figure 4.14 Path losses calculated for non-directed LOS links at every receiver location at an *xy*-plane of height 0.8 m inside the room of configuration B of Table 4.1 when the source is at the center of the ceiling.

longer distances between the source and the receiver, and longer distances should result in higher path loss. Thus, path loss is higher near the corners and lowest in the middle of the room directly underneath the source.

4.2.2   Non-Directed NLOS Links

We will discuss about impulse responses of non-directed NLOS links in this section, just as we discussed LOS links. NLOS links are formed when the radiation emitted from the source does not reach the receiver in a direct path, rather they are formed only when reflections from photons emitted from the source reach the receiver. We have seen in the previous section that reflections contribute to the increase in rms delay spread, and the smaller the contribution of the LOS response is, compared to the contribution by reflections, the larger the rms delay spread becomes. Hence, in NLOS links, since there is no LOS component in the impulse response, it is expected that the rms delay spreads will be higher than the LOS links.

For our analyses of NLOS links, we do not place the source on the ceiling usually, though we have shown in the previous section that even though the source is on the ceiling, NLOS links may be formed due to the small FOV of the receiver and large horizontal distance between the receiver and the source. Typically, when NLOS links are mentioned, it is assumed that both the source and the receiver face upward toward the ceiling. Hence, we place the source at the same height as the receiver, for example, 0.8 m in our analyses. We also first place the source and the receiver at close proximity to each other, indicating that they are probably placed on the same table or other furniture in a room. Then we place the receiver at a corner of the room, similar to previous analyses for LOS links and observe the effects on simulated impulse responses.

Figures 4.15 and 4.16 show impulse responses for two receiver locations in the room of configuration A of Table 4.1. The receiver locations are (2.5 m, 2.0 m, 0.8 m) and (0.1 m, 0.1 m, 0.8 m), respectively, while the source coordinate is (2.0 m, 2.0 m, 0.8 m). There are no LOS components in the impulse responses as expected. In Figure 4.15, the receiver and the source have a horizontal distance of 0.5 m between them. This distance is much larger in the case of Figure 4.16. Hence, we should expect larger rms delay spread in the latter case. Also, the overall magnitude of the total impulse response shown in Figure 4.15 is larger than the other one, signifying the fact that when the receiver is at the center of the room, reflections contribute more power to the total impulse response than when the receiver is at the corner. This is in contrast to the case of non-directed LOS links, where we observed the opposite behavior, that is, as the receiver moved closer to the corner, reflections contributed more power to the total impulse response.

We next show impulse responses for two receiver locations in the room of configuration B of Table 4.1 in Figures 4.17 and 4.18. The receiver locations are (3.0 m, 2.5 m, 0.8 m) and (0.1 m, 0.1 m, 0.8 m), respectively, while the source coordinate is (2.5 m, 2.5 m, 0.8 m). We have kept the horizontal distance between the source and the receiver to be 0.5 m in the case of Figure 4.17, similar to Figure 4.15. Nonetheless, we observe longer "tail" of the impulse response due to larger room size of configuration B compared to configuration A and smaller magnitude of the impulse response as the reflecting surfaces are further away from the receiver

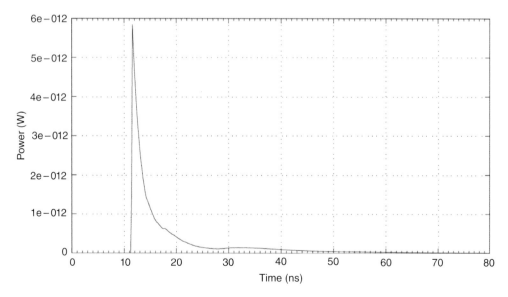

Figure 4.15 Impulse response of a non-directed NLOS link of configuration A of Table 4.1, when source coordinate is (2.0 m, 2.0 m, 0.8 m) and receiver coordinate is (2.5 m, 2.0 m, 0.8 m).

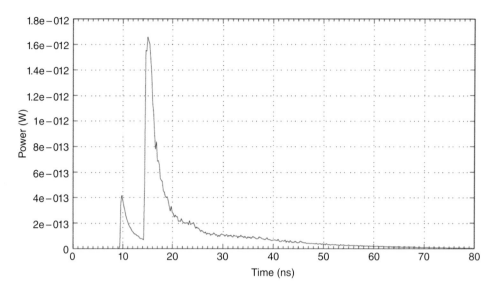

Figure 4.16 Impulse response of a non-directed NLOS link of configuration A of Table 4.1, when source coordinate is (2.0 m, 2.0 m, 0.8 m) and receiver coordinate is (0.1 m, 0.1 m, 0.8 m).

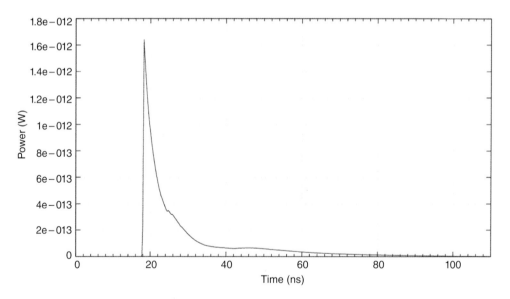

Figure 4.17 Impulse response of a non-directed NLOS link of configuration B of Table 4.1, when source coordinate is (2.5 m, 2.5 m, 0.8 m) and receiver coordinate is (3.0 m, 2.5 m, 0.8 m).

and hence the reflected optical signal experiences more path loss than that of configuration A. The overall shapes of the impulse response are unchanged compared to the impulse responses of configuration A.

Finally, we show impulse responses for two receiver locations in the room of configuration C of Table 4.1 in Figures 4.19 and 4.20. The receiver locations are (3.75 m, 2.75 m, 0.8 m) and

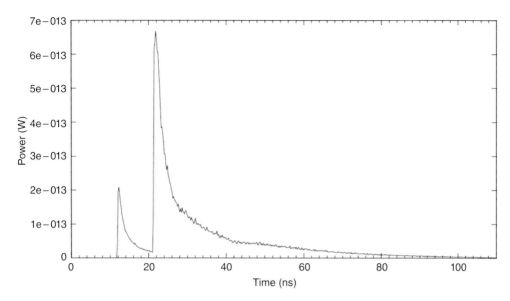

Figure 4.18 Impulse response of a non-directed NLOS link of configuration B of Table 4.1, when source coordinate is (2.5 m, 2.5 m, 0.8 m) and receiver coordinate is (0.1 m, 0.1 m, 0.8 m).

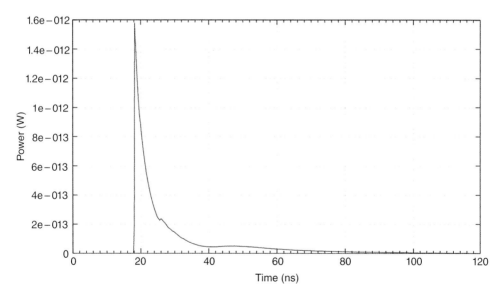

Figure 4.19 Impulse response of a non-directed NLOS link of configuration C of Table 4.1, when source coordinate is (3.25 m, 2.75 m, 0.8 m) and receiver coordinate is (3.75 m, 2.75 m, 0.8 m).

(0.1 m, 0.1 m, 0.8 m), respectively, while the source coordinate is (3.25 m, 2.75 m, 0.8 m). Similar to other configurations, we observe diminishing contribution of reflections to the total impulse response as the receiver moves toward the corner. The overall magnitude of the impulse responses of Figures 4.19 and 4.20 is also smaller than previous configurations, as

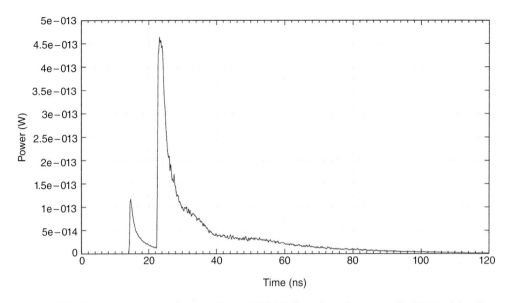

Figure 4.20   Impulse response of a non-directed NLOS link of configuration C of Table 4.1, when source coordinate is (3.25 m, 2.75 m, 0.8 m) and receiver coordinate is (0.1 m, 0.1 m, 0.8 m).

larger room size implies larger distances between the receiver and the reflecting surfaces, and hence larger path loss for the reflected optical signal. Again, the overall shapes of the impulse responses are unchanged compared to the other two configurations.

We will now show the magnitude responses of configurations A and C calculated from these impulse responses. As shown in Figure 4.21, all the magnitude responses decrease very sharply since there is no LOS component in the impulse responses. We have tabulated the measured −3 dB bandwidth in Table 4.5 for each of the channels shown in Figure 4.21. As expected, because of larger room size, −3 dB bandwidths of the NLOS channels of configuration C are smaller than those of configuration A.

We have shown in Table 4.6 the power contributions from 3rd to 10th order reflections and the rms delay spreads of each of the NLOS channels of configurations A and C of Table 4.1. Similar to the −3 dB bandwidths shown, configurations A and C exhibit expected behavior where the values increase as room size increases.

Hence, from these discussions, we can conclude that, for NLOS channels, the similar trend of increasing rms delay spreads and decreasing −3 dB bandwidths can be expected for larger room sizes.

We will now proceed to show rms delay spreads of NLOS links at all receiver locations, as we did in the case of LOS links, inside the room of configuration B when the source is placed at the center of the room. Figure 4.22 shows rms delay spreads as the receiver is placed at different locations. From our analyses, it is evident that as the receiver moves toward the corner, rms delay spread will increase and this is what we observe from the figure.

We will show how receiver FOV affects rms delay spreads of NLOS channels. For this purpose, we have chosen to observe the effects in the room of configuration C. The source location is fixed at (6.0 m, 2.75 m, 0.8 m). We place the receiver at (3.75 m, 2.75 m, 0.8 m),

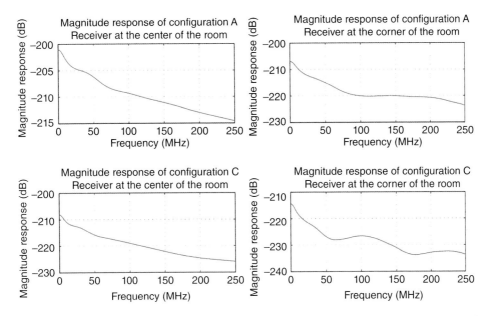

Figure 4.21 Magnitude responses calculated from impulse responses of NLOS channels simulated for two locations of the receiver for configurations A and C of Table 4.1.

Table 4.5 −3 dB bandwidth of NLOS channels consisting of two locations of the receiver for configurations A and C of Table 4.1

	Receiver at the center of the room	
	Configuration A	Configuration C
−3 dB Bandwidth	17.13 MHz	10.57 MHz
	Receiver at the corner of the room	
	Configuration A	Configuration C
−3 dB Bandwidth	12.04 MHz	8.52 MHz

Table 4.6 Percentage of total power contributed by 3rd to 10th order reflections and rms delay spread calculated from impulse responses of non-directed NLOS links of configurations A and C of Table 4.1 at two receiver positions

		Configuration A	Configuration C
Percentage of total power contributed by 3rd to 10th order reflections	Receiver at the center	18.2233%	19.9186%
	Receiver at the corner	38.4146%	41.0853%
Root-mean-square delay spread	Receiver at the center	2.7179 ns	4.4771 ns
	Receiver at the corner	4.2596 ns	6.6862 ns

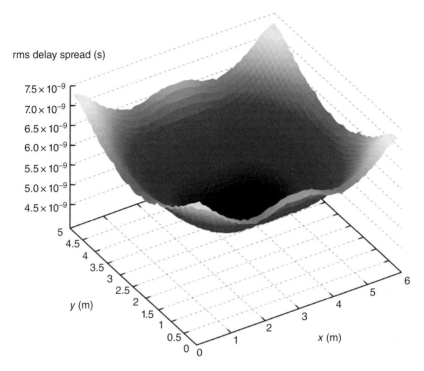

Figure 4.22 rms delay spreads calculated for non-directed NLOS links at every receiver location at an *xy*-plane of height 0.8 m inside the room of configuration B of Table 4.1 when the source coordinate is (3.0 m, 2.5 m, 0.8 m) and the source emits toward the ceiling.

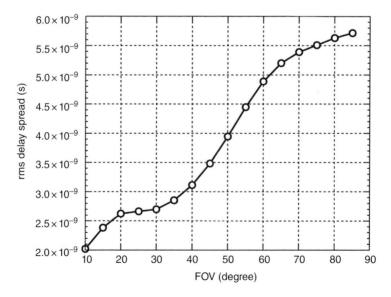

Figure 4.23 Effect of receiver FOV on rms delay spread of a non-directed NLOS link when source coordinate is (6.0 m, 2.75 m, 0.8 m) and receiver coordinate is (3.75 m, 2.75 m, 0.8 m) in the room of configuration C of Table 4.1.

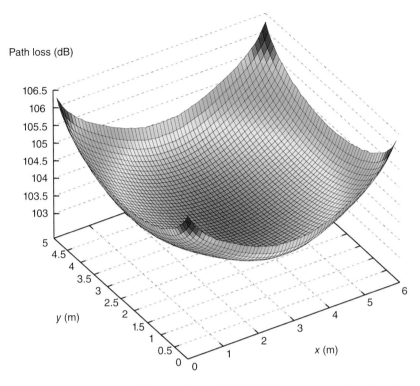

Figure 4.24 Path losses calculated for non-directed NLOS links at every receiver location at an *xy*-plane of height 0.8 m inside the room of configuration B of Table 4.1 when the source coordinate is (3.0 m, 2.5 m, 0.8 m) and the source emits toward the ceiling.

where we change the FOV from 10° to 85° in an increment of 5° at each location and plot the calculated rms delay spread from the simulated impulse responses. Figure 4.23 shows the change in rms delay spread as the FOV of the receiver is changed. We do not observe anything that is unexpected in this scenario. As receiver FOV increases, more light is captured from reflections, and hence rms delay spread also increases.

Similar to our calculations of path losses of non-directed LOS links, we will now repeat the procedure for non-directed NLOS links as well. Higher path loss usually implies longer distance from the source to the receiver, and hence, similar to the LOS case, we should see higher path loss at room corners. Figure 4.24 shows path losses for every receiver location at an *xy*-plane of 0.8 m in the room of configuration B of Table 4.1 when the source is located at (3.0 m, 2.5 m, 0.8 m). As we can see from the figure, our results are similar in nature to those obtained and shown in Figure 4.22, that is, path losses are indeed larger at room corners. Hence, we can conclude that non-directed LOS and NLOS links both behave in a similar manner regarding path losses.

4.3 Effects of Furniture on Root-Mean-Square Delay Spread

Until now we have focused solely on analyzing impulse responses of configurations where the room was empty, that is, devoid of any furniture. This has kept our analyses simple and showed that the principal factors in determining the rms delay spread of an indoor OWC includes the dimension of the room, the locations of the source, and the receiver and the receiver's FOV. These findings are valid; however, in real-world scenarios, no room is actually devoid of furniture, and hence some consideration should be given to the possibility of the presence of furniture in a room and calculation of rms delay spread should be carried out to see the effects of furniture on rms delay spread, if any. Our focus in this section is to simulate such a condition where some furniture is present.

In the previous sections, we have carried out rms delay spread calculations of every receiver location in the room of configuration B of Table 4.1 for both non-directed LOS and NLOS links. To compare the change in rms delay spreads, we will hence introduce some furniture in the same room configuration and conduct our calculations.

The room of configuration B of Table 4.1 is a mid-size office room, and hence we have introduced some tables and lockers in the room. Figure 4.25 shows their positions clearly.

We have placed four work tables and two lockers at the two corners of the room. All the furniture have been modeled as boxes, which simplifies impulse response simulations. Each furniture can be specified by a reflection coefficient and two coordinates. The two coordinates of the furniture, as shown in Figure 4.25, are coordinates of the lower right corner of the front face (x_0, y_0, z_0) and the upper left corner of the back face (x_1, y_1, z_1). With these two coordinates, it is possible to accurately define the dimension, that is, the length, the width, and the height of the furniture. In Table 4.7, we have specified the two coordinates of the furniture and their associated reflection coefficients.

We have conducted impulse response simulations for non-directed LOS links where we placed the source at the center of the ceiling and the receiver at every location in an xy-plane of height 0.8 m in the room. Since there is now furniture in the room, we obviously cannot place the receiver at all locations, especially as the two lockers at the two corners have a higher height. Hence, excluding those two areas which the two lockers occupy, we have carried out calculations of rms delay spreads. Now, from Figure 4.8, we already have previous results of rms delay spreads when there was no furniture. Hence it is possible for us to compare the two

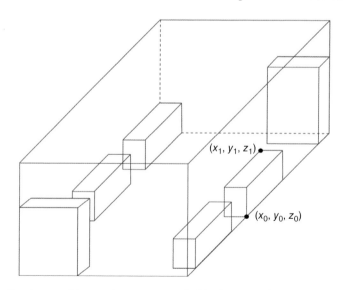

Figure 4.25   Furniture positions and arrangements inside the room of configuration B of Table 4.1.

Table 4.7   Coordinates and reflection coefficients of furniture in the room of configuration B of Table 4.1

	Coordinate 1 (x_0, y_0, z_0)	Coordinate 2 (x_1, y_1, z_1)	Reflection coefficient
Furniture 1	(0.5 m, 0 m, 0 m)	(2.0 m, 0.6 m, 0.8 m)	0.5
Furniture 2	(2.5 m, 0 m, 0 m)	(4.0 m, 0.6 m, 0.8 m)	0.6
Furniture 3	(5.5 m, 0 m, 0 m)	(6.0 m, 1.2 m, 2 m)	0.7
Furniture 4	(4.5 m, 4.4 m, 0 m)	(6.0 m, 5.0 m, 0.8 m)	0.5
Furniture 5	(2.5 m, 4.4 m, 0 m)	(4.0 m, 5.0 m, 0.8 m)	0.5
Furniture 6	(0 m, 3.5 m, 0 m)	(0.6 m, 5 m, 1.8 m)	0.85

situations. Since there is furniture now in the room, and furniture means more reflecting surfaces, it is expected that the rms delay spread will increase somewhat. We have thus calculated the percentage increase in rms delay spreads of all locations of the receiver in an xy-plane of height 0.8 m by the following:

$$\text{Percentage increase in rms delay spread} =$$
$$\frac{\text{rms delay spread with furniture} - \text{rms delay spread without furniture}}{\text{rms delay spread without furniture}} \times 100\% \quad (4.4)$$

We have shown our results in Figure 4.26. The figure is the top view of the room, where we show the result calculated using (4.4) as a color. The color bar at the right side of the figure is the legend that explains the values represented by the colors. We see that for most of the areas of the room, the percentage increase in rms delay spread is less than approximately 10%. We see a significant increase in rms delay spread near the two lockers of the room, however. The percentage increase quickly jumps approximately from 15 to 45% as the receiver moves closer to the lockers. This behavior is not unexpected, however, and we can explain it easily by studying the parameters given in Table 4.7. The two lockers, furniture 3 and furniture 6 in Table 4.7, have high reflection coefficients, assuming they are colored in white. The other furniture in the room have relatively smaller reflection coefficients assuming they are colored darker than these two lockers. Moreover, the two lockers have a height of 2.0 m and 1.8 m, respectively. Hence, reflections from their surfaces contribute much higher power to the total impulse response when the receiver is near them. This is the reason why we observe a larger percentage increase in rms delay spread as the receiver moves closer to the lockers.

We will conduct the same simulations using non-directed NLOS links. In Figure 4.22, we have shown rms delay spreads for all receiver locations at an xy-plane of height 0.8 m when the source coordinate was (3.0 m, 2.5 m, 0.8 m) in the room of configuration B of Table 4.1. We can simulate impulse responses for the same receiver locations and in the same configuration, except putting furniture as shown in Figure 4.25. Hence, similar to our LOS link analyses, we can show percentage increase in rms delay spread for addition of furniture in the room. Figure 4.27 shows the results from our calculations. We see very similar trends to what we observed in Figure 4.26. Rms delay spread increases the most near the two lockers of the room, while at other places the increase is quite low. Even compared to the LOS case, we see that the highest percentage increase is around approximately 17%, where in the former case, it was around approximately 47%. This is due to the reason that rays from the source hit the

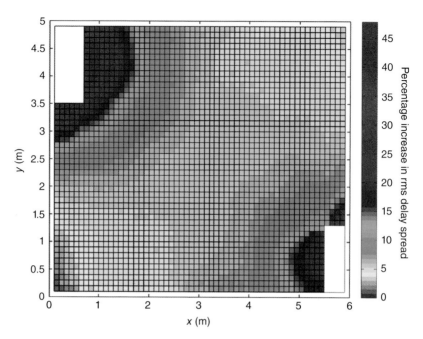

Figure 4.26 Percentage increase in rms delay spreads in a non-directed LOS link in the room of configuration B of Table 4.1 with furniture, compared to without furniture, when the source is at the center of the ceiling and the height of the receiver is 0.8 m.

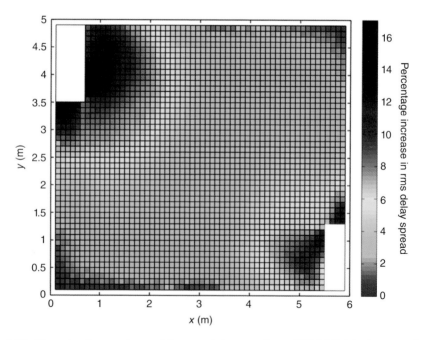

Figure 4.27 Percentage increase in rms delay spreads in a non-directed NLOS link in the room of configuration B of Table 4.1 with furniture, compared to without furniture, when the height of the receiver is 0.8 m and source coordinate is (3.0 m, 2.5 m, 0.8 m).

ceiling first in case of non-directed NLOS links when the source is facing directly upward toward the ceiling, and so, furniture placed inside the room only receives light from reflections. As a result, reflections from furniture do not have as much power as those in non-directed LOS links because in case of non-directed LOS links, it is certainly possible that rays from the source directly hit the furniture before undergoing any reflections. Hence, we can draw the conclusion that rms delay spread is less affected with furniture in non-directed NLOS links compared to non-directed LOS links.

From our discussions on the effects of furniture on rms delay spread, we found that indeed addition of furniture increases rms delay spread somewhat; however, the most significant increase is observed closer to furniture that is placed at a more height than the receiver. Also we have seen that non-directed LOS links are affected more than non-directed NLOS links. Still, for most of the areas of the room, we found that percentage increase in rms delay spread is less than ~10% for non-directed LOS links and less than ~5% for non-directed NLOS links. Hence, the conclusion can be drawn that performance results computed from impulse responses of an empty room will also be approximately correct in real-world scenarios where some furniture will always be present. Unless the receiver is located close to some furniture higher than the receiver itself, impulse response simulations in an empty room will not lead to grossly wrong results when performance is calculated. Thus, due to the simplicity in calculating impulse responses of an empty room, we will in general always use an empty room as a model when finding impulse responses of some particular room configuration.

4.4   SNR Calculations and BER Performance

Knowing the rms delay spread of an indoor OWC enables us to calculate the maximum bit rate that is possible to transmit without ISI and without equalization at the receiver side. However, for measuring BER performance of a system, knowledge of only the rms delay spread will not be enough. The modulation method used for intensity-modulation direct-detection (IM/DD) systems usually is chosen such that bandwidth efficiency is achieved while sacrificing some power efficiency. Since indoor OWCs are band-limited due to multipath reflections, power efficiency is sacrificed assuming the transmitter will have enough power, but bandwidth efficiency is given priority to transmit at as high data rate as possible. The main schemes of single-carrier modulation formats in the IM/DD systems include M-PAM (Pulse Amplitude Modulation) and L-PPM (Pulse Position Modulation). Though other modulation schemes are possible whose baseband output is real, as optical communications require real baseband signals explained in Chapter 2, these two schemes are widely studied. It is known [5–7] that PAM systems are more bandwidth efficient than PPM schemes while PPM modulation is more power efficient. Hence, usually PAM is employed as the choice of modulation schemes, especially considering simplicity, 2-PAM or, in other words, OOK (on-off keying) is the popular choice for optical wireless systems. We will also analyze our systems in this chapter assuming data is transmitted using OOK modulation format.

The probability of error is dependent on SNR as given by

$$P_e = Q\left(\sqrt{\text{SNR}}\right) \tag{4.5}$$

where $Q(x)$ is given by $Q(x) = 1/2\,\text{erfc}\left(x/\sqrt{2}\right) = 1/\sqrt{2\pi}\int\limits_{x}^{\infty}\exp\left(-t^2/2\right)dt$ [8].

The SNR for indoor optical wireless systems can be calculated by

$$\text{SNR} = \left(\frac{\mathcal{R} \times (P_{s1} - P_{s0})}{\sigma_t} \right)^2 \tag{4.6}$$

where $\mathcal{R}$ is the photodiode responsivity (A/W), and P_{s1} and P_{s0} represent the received optical power associated with logic 1 and logic 0, respectively. Hence, by specifying $(P_{s1} - P_{s0})$, actually the eye opening from the eye diagram at the sampling instant is specified, that is, this takes multipath-induced ISI into consideration.

The noise variance σ_t^2 consists of noise associated with logic 1 and logic 0, that is, $\sigma_t = \sigma_0 + \sigma_1$. Both of these quantities can be divided into three components [9], namely $\sigma_0^2 = \sigma_{bn}^2 + \sigma_{pr}^2 + \sigma_{s0}^2$ and $\sigma_1^2 = \sigma_{bn}^2 + \sigma_{pr}^2 + \sigma_{s1}^2$. Here σ_{bn}^2 denotes the background light–induced shot noise variance and can be calculated by

$$\sigma_{bn}^2 = 2q I_2 p_{bn} A_R \Delta\lambda \mathcal{R} B \tag{4.7}$$

where q is the charge of an electron, p_{bn} is the background light irradiance per unit wavelength on the photodiode, A_R is the detection area of the photodiode, $\Delta\lambda$ is the bandwidth of the optical filter which is necessary to use to reduce background light, $\mathcal{R}$ is the photodiode responsivity, B is the bit-rate of the system, and I_2 is a noise bandwidth factor and is a function of the transmitter pulse shape and equalized pulse shape only, and is independent of bit rate [10–13].

The next noise component is σ_{pr}^2, which denotes the preamplifier-induced noise variance. Preamplifiers are usually field-effect-transistor (FET) trans-impedance preamplifiers where noise sources include Johnson noise associated with the FET channel conductance, Johnson noise from the load or feedback resistor, shot noise arising from gate leakage current and $1/f$ noise [14]. All these components are summed up by the following expression:

$$\sigma_{pr}^2 = \left(\frac{4kT}{R_F} + 2q I_L \right) I_2 B + \frac{4kT\Gamma}{g_m} (2\pi C_T)^2 A_F f_c B^2 + \frac{4kT\Gamma}{g_m} (2\pi C_T)^2 I_3 B^3 \tag{4.8}$$

where k is the Boltzmann's constant, T is the absolute temperature, q is the electron charge, B is the electrical bandwidth, R_F is the feedback resistance, I_L is the total leakage current that includes FET gate current and dark current of the photodiode, I_2 and I_3 are weighting functions that are dependent only on the input optical pulse shape to the receiver and the equalized output pulse shape, Γ is a noise factor associated with channel thermal noise and gate-induced noise in the FET, g_m is the FET trans-conductance, C_T is the total input capacitance consisting of photodiode and stray capacitance, A_F is the weighting function where $A_F = 0.184$ for non-return-to-zero (NRZ) coding format and f_c is the $1/f$ corner frequency of the FET. For simplicity, the FET gate leakage current and $1/f$ noise are usually neglected [14]. Hence, σ_{pr}^2 can be given by

$$\sigma_{pr}^2 = \frac{4kT}{R_F} I_2 B + \frac{4KT\Gamma}{g_m} \left(2\pi C_T\right)^2 I_3 B^3 \tag{4.9}$$

Shot noise induced by the received signal which consists of shot noise σ_{s1}^2 when a logic 1 is received and shot noise σ_{s0}^2 when a logic 0 is received are very small and can be neglected. Hence the SNR can be expressed as

$$SNR = \left(\frac{\mathcal{R} \times \left(P_{s1} - P_{s0}\right)}{2\sqrt{\sigma_{bn}^2 + \sigma_{pr}^2}}\right)^2 \tag{4.10}$$

Let us put some realistic values into (4.10) to find out the required received power level for a given SNR. We will use an avalanche photodiode (APD) in this case as simple p-i-n photodiodes do not have enough gain to work in a diffused system without complex optics. An APD requires an excess noise factor of $\sqrt{F}$ that is multiplied with the shot noise expression of (4.7). Typically for Si APDs, the value of F is around 4. Hence, for σ_{bn}^2, we will use $q=1.6\times10^{-19}$ C, $p_{bn}=5.8\,\mu$W/cm^2.nm according to [7], $I_2=0.562$ for rectangular pulse shapes, $A_R=1\times10^{-6}$ m^2, $\Delta\lambda=30$ nm, $\mathcal{R}=50$ A / W (assuming an avalanche photodetector) and $B=500$ Mbps as the data rate. For σ_{pr}^2, we will use $k=1.38\times10^{-23}$ J K^{-1}, $T=300$ K, $I_2=0.562$ and $I_3=0.0868$, $\Gamma=0.82$, $R_F=1.4$ kΩ, $g_m=14$ mS, $C_T=2.0$ pF. For a BER of 10^{-6}, the required SNR from (4.5) is $\left(Q^{-1}\left(P_e\right)\right)^2 = 22.595$. Hence, by using these values in (4.10), we obtain $\left(P_{s1}-P_{s0}\right) = 0.0269\,\mu$W.

Hence, the factor to consider is whether this amount of power can be received by the photodiode or not. We have used a lower bandwidth photodiode to increase its detection area than what we have used in our simulations in the previous sections, and the APD also gives much higher responsivity. For example, when the source is at the center of the ceiling in the room of configuration B in Table 4.1, we have simulated the impulse response for non-directed LOS links of every receiver location at an xy-plane of height 0.8 m. The LOS response is 0.0437 μW at the location when the receiver is in the middle of the room and when the area of the receiver is 1×10^{-6} m^2 assuming source transmit power is 1 W. This is sufficient for obtaining a BER performance of 10^{-6} as explained above.

But at other points of the room, the LOS response becomes smaller and multipath effects come into play that reduce the received power difference $(P_{s1}-P_{s0})$. Moreover, the output power of the source is also needed to be limited to much lower values than 1 W considering eye-safety issues related to infrared radiation. Hence, even non-directed LOS links, let alone non-directed NLOS links, using real parameters of photodiodes that are available off-the-shelf, will not achieve any acceptable BER performance limit. The solution is to increase the collected power by using a focusing lens in front of the photodiode. A properly designed system where the photodiode is at the focal point of a focusing lens will capture much more optical power. For example, if we use a lens having a diameter of 2.5 cm, the area of the lens will be 4.9087×10^{-4} m^2. Hence, the increase in received power compared to the bare photodiode of area 1×10^{-6} m^2 in the previous example will be 490.87 or 26.91 dB.

Let us check the worst case multipath scenario of the room we simulated earlier. We found the worst case when the receiver is at the corner of the room, that is, (0.1 m, 0.1 m, 0.8 m). Now we can obtain the eye diagram at that location by convolving rectangular pulses generated for random bit sequences with the impulse response and see the effects of multipath-induced ISI.

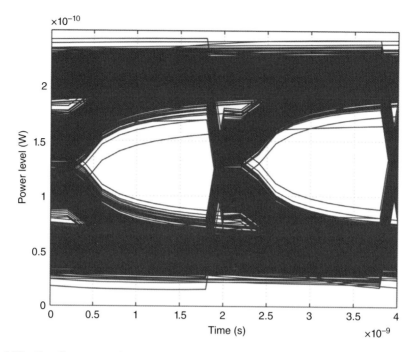

Figure 4.28 Eye diagram considering rectangular pulse shaping at a bit rate of 500 Mbps at receiver coordinate of (0.1 m, 0.1 m, 0.8 m) in a non-directed LOS link of configuration B of Table 4.1 when the source is at the center of the ceiling.

The peak power of logic 1 is assumed to be 20 mW while for logic 0 it is assumed that the laser is turned off. Figure 4.28 shows the eye diagram at the farthest corner of the room, considering only the impulse response of that location and rectangular pulse shaping with a bit rate of 500 Mbps.

From this figure, we see that eye opening at the sampling instant is only $1.752 \times 10^{-10} - 8.321 \times 10^{-11}$ W $= 9.2 \times 10^{-11}$ W, which is far below than what is required for a BER of 10^{-6} as calculated earlier. However, by using a focusing lens as specified above, we can add 26.91 dB to this value and the end result is 0.045 μW, which is more than sufficient to ensure the specified BER.

4.5 Impact of Higher Order Reflections

In this section we will analyze if higher order reflections are necessary for evaluating some performance metrics such as rms delay spread. As discussed in the previous chapter, if Barry's algorithm is used, it may not be always possible, due to unavailability of computing time or resource to calculate impulse responses having contributions of fourth or higher order reflections. Of course, though using the MMC or the CDMMC algorithm enables one to calculate impulse responses having contributions of as many orders of reflections as necessary quite easily, it is still relevant to discuss this as currently many research literature concludes on performance metrics based on impulse responses having only up to third-order reflections.

We will introduce a multi-spot diffusing architecture in this section, whereas there are nine diffuse spots on the ceiling, that is, there are nine sources on the ceiling. These sources can be created from a single laser source by placing a holographic diffuser in front of it. The holographic diffuser can be designed such that the single laser beam splits into nine beams and thereby nine sources are created as they hit the ceiling. We, at first, choose three typical test locations to analyze the contribution of each order of reflections to the total impulse response. The three locations are as follows: A (0 m, 0 m, 0.9 m) representing a point at the room corner, where severe multipath reflections are experienced; B (0 m, 3 m, 0.9 m) representing a point near a wall but away from corners, where medium multipath reflections are experienced; and C (3 m, 3 m, 0.9 m) representing a point at the center of the room, where weak multipath reflections are experienced. Next, we demonstrate the estimation accuracy distributions all over the room. Other parameters of the room, the sources, and the receiver are given in Table 4.8.

Figures 4.29, 4.30, and 4.31 show the contributions of each order of reflections to the impulse responses at test locations A, B, and C, respectively. The total impulse response is the sum of all these contributions. In all three locations, the 0th reflection (LOS) contributes the most significant amount of optical power. From the third-order reflections, the shapes of individual impulse responses are similar; nevertheless, they attenuate and temporally spread out as the order increases. Compared with low-order reflections, high-order reflections contribute less power but more delay to the total impulse response. Unlike received power, which is only related to impulse response amplitude, delay spread is jointly determined by impulse response amplitude and delay. High-order reflections, therefore, obviously make a more significant impact on delay spread than received power. In other words, delay spread estimation should converge slower than power as reflection orders increase.

To further explore the convergence differences, we plot the estimation accuracy of received optical power, received signal power, average delay, and rms delay spread in Figures 4.32, 4.33, and 4.34. The results are calculated by first applying 20 orders of reflections as references of accurate estimation. The calculated accuracies are defined as the ratio of estimated values to accurate values that were calculated by including 20 orders of reflections.

Table 4.8 Simulation parameters for analyzing impact of higher order reflections

Room size: length × width × height	6 m × 6 m × 3 m
Laser source location	(3 m, 3 m, 0.5 m)
Locations of diffuse transmitting spots	(1.5 m, 1.5 m, 3 m) (1.5 m, 3 m, 3 m) (1.5 m, 4.5 m, 3 m)
	(3 m, 1.5 m, 3 m) (3 m, 3 m, 3 m) (3 m, 4.5 m, 3 m)
	(4.5 m, 1.5 m, 3 m) (4.5 m, 3 m, 3 m) (4.5 m, 4.5 m, 3 m)
Transmitted optical power at each spot	1 W
Reflection coefficients (ceiling, wall, floor)	0.9, 0.7, 0.1
Reflection element size	0.2 m × 0.2 m
Receiver FOV	$60°$
Receiver aperture area	$1 \times 10^{-4}\,\text{m}^2$
Time resolution	0.66 ns

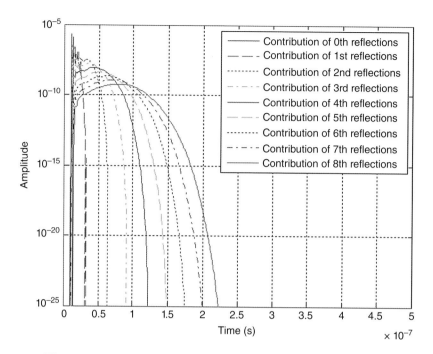

Figure 4.29 Impulse response of each order of reflections at location 1.

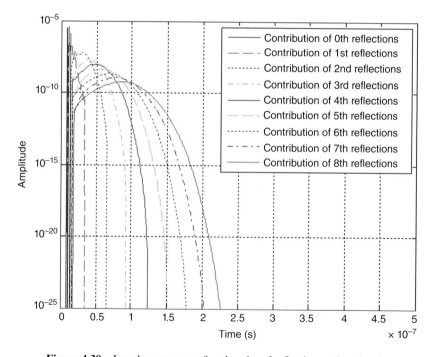

Figure 4.30 Impulse response of each order of reflections at location 2.

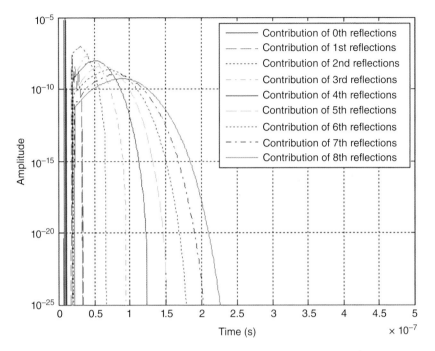

Figure 4.31 Impulse response of each order of reflections at location 3.

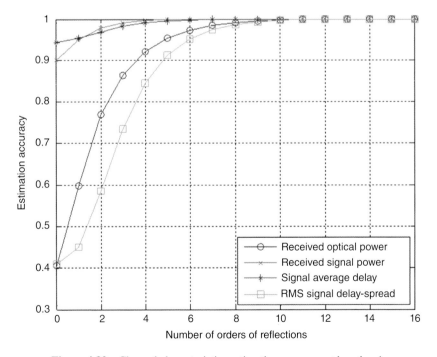

Figure 4.32 Channel characteristics estimation accuracy at location 1.

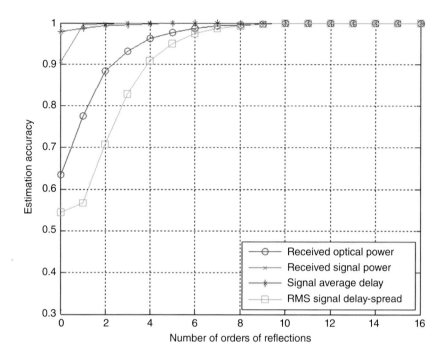

Figure 4.33   Channel characteristics estimation accuracy at location 2.

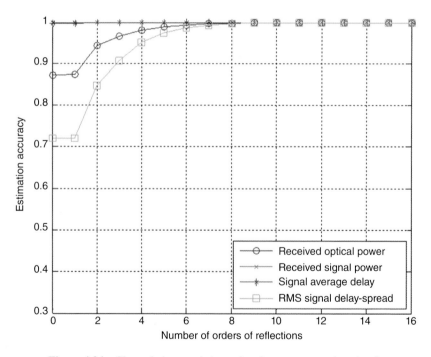

Figure 4.34   Channel characteristics estimation accuracy at location 3.

The figures show that signal power and average delay converge fast and will not be noticeably impacted by truncating them to the first two or three orders; however, the received optical power and the rms delay spread converge substantially slower. For a Gbps high-speed transmission system, delay spread deserves particular attention, because it dominantly determines ISI. The maximum delay spread observed all over the room is 3.66 ns. To make sure any delay spread model error is smaller than half of a symbol period, we need the delay spread estimation accuracy higher than $1-(1\times10^{-9}/(2\times3.66\times10^{-9}))\times100\%=86.3\%$ for 1 Gbps systems and $1-(1\times10^{-10}/(2\times3.66\times10^{-9}))\times100\%=98.6\%$ for 10 Gbps systems. As the figures show, estimation accuracies for the three locations by first three orders are only 73.4%, 82.9%, and 90.7%, respectively. If we only use first three orders of reflections to create the model, only location C meets the accuracy needed for 1 Gbps systems and none of them satisfy 10 Gbps systems.

Now, as indoor OWC systems are expected to provide full coverage and mobility, it is necessary to extend channel model analysis from the three test points to the entire area of the room. It can be shown by extending the results from earlier discussions that high-order reflections make impacts differently at different locations. Here, the spatial distribution is explored by simulating 841 channels, representing every piece of 0.2 m × 0.2 m area of a 6 m × 6 m × 3 m room. As we demonstrated that delay spread experiences the most severe impact from discarded high-order reflections, we utilize delay spread estimation accuracy to indicate model accuracy as the worst case. The contours for each additional order of reflections considered are shown in Figures 4.35, 4.36, 4.37, 4.38, 4.39, 4.40, 4.41, and 4.42, respectively. Generally, the shapes of the contours are similar as the number of reflections increases: high accuracy

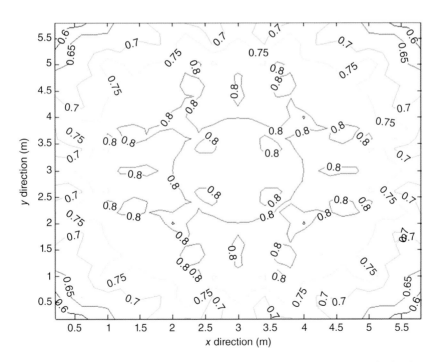

Figure 4.35 Delay-spread model accuracy contour considering first two orders of reflections (%).

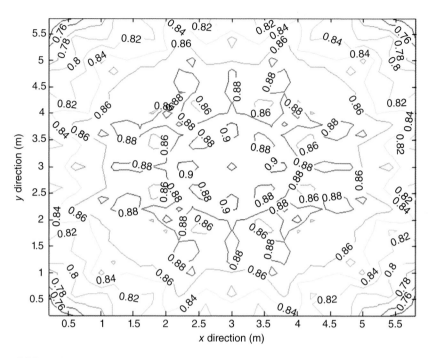

Figure 4.36   Delay-spread model accuracy contour considering first three orders of reflections (%).

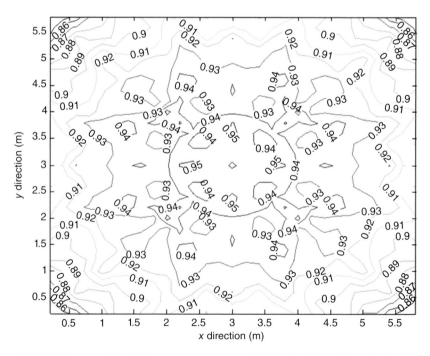

Figure 4.37   Delay-spread model accuracy contour considering first four orders of reflections (%).

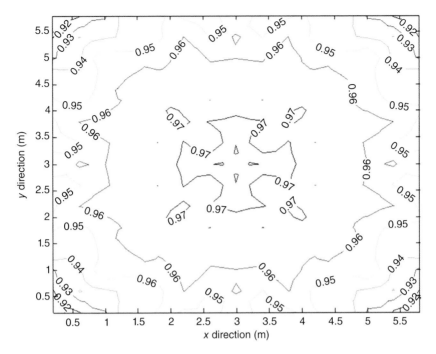

Figure 4.38 Delay-spread model accuracy contour considering first five orders of reflections (%).

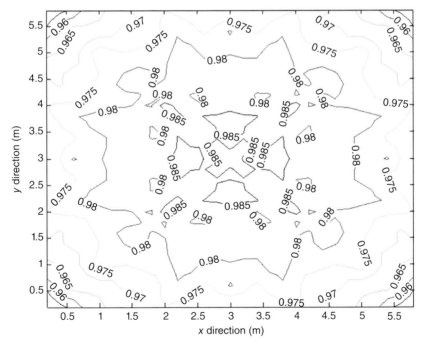

Figure 4.39 Delay-spread model accuracy contour considering first six orders of reflections (%).

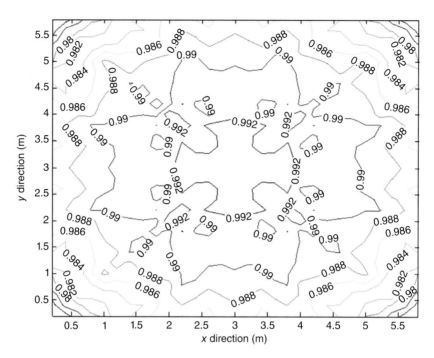

Figure 4.40   Delay-spread model accuracy contour considering first seven orders of reflections (%).

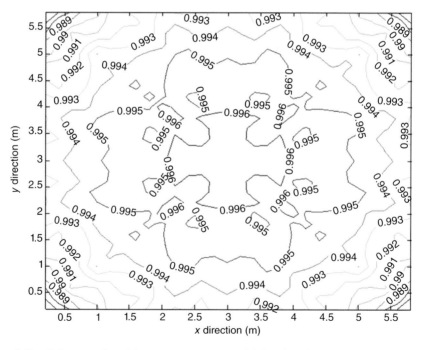

Figure 4.41   Delay-spread model accuracy contour considering first eight orders of reflections (%).

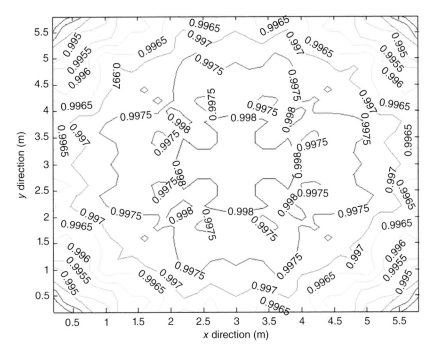

Figure 4.42 Delay-spread model accuracy contour considering first nine orders of reflections (%).

areas are near the center of the room and the accuracy decreases when approaching the corners of the room, where multipath effect is much higher; we also observe that dense contours exist at the corner of the room, which indicates a sharp decrease in model accuracy. We are able to get the required reflection orders for specific transmission rate from the contours. For instance, to ensure the accuracy of the entire room is above the need for 1 Gbps data rate, we need 5 orders (86.3% above accuracy guaranteed) of reflections and 9 orders (98.6% above accuracy guaranteed) for 10 Gbps systems.

As discussed in the last chapter, if deterministic algorithms are used for simulating impulse responses such as Barry's algorithm or its variations, computational complexity increases considerably with the number of reflection orders included. Hence, it is reasonable that many researchers using deterministic algorithms neglect high-order reflections to be able to conduct simulations in a practical period of time. However, this sacrifice of accuracy for efficiency results in more and more significant performance errors. A reasonable and feasible solution to keep both accuracy and efficiency is applying calibration. We discuss here a calibration method based on the statistical data of the accuracy curves. The accuracy curves of all 841 locations are drawn in Figure 4.43. By averaging them, one can obtain the general calibration curve $c[k]$. It provides the correction value for each order of reflections. For the delay spread calculated from first k orders of reflections, the value can be calibrated by

$$\tilde{s}_j^{(k)} = \frac{s_j^{(k)}}{c[k]} \tag{4.11}$$

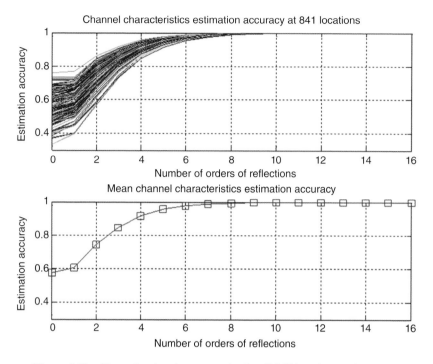

Figure 4.43   Channel estimation accuracies for all 841 locations and average.

Table 4.9   Calibration values for channel model

Orders	1	2	3	4	5	6
Calibration (%)	60.53	74.53	84.71	91.77	95.52	97.67

Orders	7	8	9	10	11	12
Calibration (%)	98.79	99.38	99.69	99.84	99.92	99.96

where $\tilde{s}_j^{(k)}$ and $s_j^{(k)}$ are the post-calibrated and pre-calibrated delay spreads from the first k orders of reflections, respectively.

The calibration value for each order of reflections is given in Table 4.9.

We apply the rms error model to compare the performance before and after calibration as in Figure 4.44. These are calculated by (4.12) and (4.13), respectively:

$$e^{(k)} = \frac{1}{N} \sum_{j=1}^{N} \left(\frac{s_j^{(k)} - s_j}{s_j} \right)^2 \tag{4.12}$$

$$\tilde{e}^{(k)} = \frac{1}{N} \sum_{j=1}^{N} \left(\frac{\tilde{s}_j^{(k)} - s_j}{s_j} \right)^2 \tag{4.13}$$

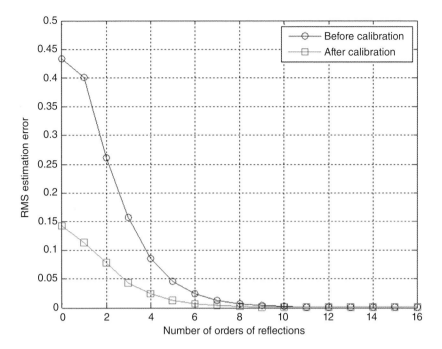

Figure 4.44   Comparison of average rms error before and after calibration.

where N is the total number of channels tested, $e^{(k)}$ is the pre-calibrated estimation error, $\tilde{e}^{(k)}$ is the post-calibrated estimation error, and s_j is the reference of accurate estimation. As we can see, there is a substantial decrease in delay spread error after calibration. From Figure 4.44, after applying first three orders of reflections, the average rms error drops from 15.7 to 4.3%. We draw the contour for the calibrated model accuracy in Figures 4.45 and 4.46. It can be observed that the order of reflections needed for 1 Gbps systems reduces from 5 to 3, and for 10 Gbps systems the needed order of reflections reduces from 9 to 7.

4.6   Conclusions

In this chapter we have conducted detailed analyses on impulse responses of various room configurations and shown examples of specific receiver locations as necessary. We have seen that LOS channels have better -3 dB bandwidth and smaller rms delay spread compared to NLOS channels. Hence, to achieve a high data rate, an optical wireless system should be designed such that an LOS link exists between the source and the receiver. We have also shown effects of adding furniture in a room. The conclusion is that if the receiver location is not near some furniture which is at a higher height than the receiver itself, the results obtained from analyzing an empty room can be used interchangeably without introducing too many significant errors. Lastly, we have conducted SNR calculations and shown that using a bare photodiode, even an APD, will not be good enough to obtain

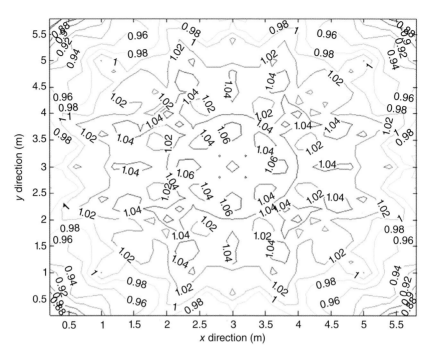

Figure 4.45 Calibrated delay spread model accuracy contour considering first three orders of reflections (%).

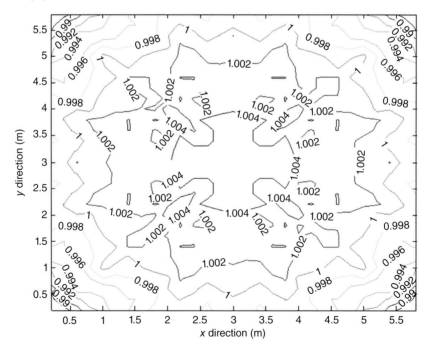

Figure 4.46 Calibrated delay spread model accuracy contour considering first seven orders of reflections (%).

sufficient SNR to ensure a BER of 10^{-6}, supposing forward-error-correction (FEC) is employed at the receiver side. The solution is to use a focusing lens that significantly increases the received power and hence SNR becomes high enough for obtaining a good BER performance from the system.

References

[1] Menlo Systems, "APD310 – High Sensitivity Detector Unit," Datasheet of High Sensitivity Detector Unit APD210/310, January 2009.

[2] Thorlabs, Inc., "FPD310-FV – Freespace High Sensitivity PIN Photo Detector Unit," Datasheet of Freespace High Sensitivity PIN Photo Detector Unit, June 2007.

[3] A. Goldsmith, *Wireless Communications*. Cambridge University Press, Cambridge, 2005.

[4] T. S. Rappaport, *Wireless Communications: Principles and Practice*, 2nd ed. Prentice Hall, Upper Saddle River, NJ, 2002.

[5] T. Y. Elganimi, "Performance comparison between OOK, PPM and PAM modulation schemes for free space optical (FSO) communication systems: analytical study," *International Journal of Computers and Applications*, vol. 79, no. 11, pp. 22–27, 2013.

[6] J. M. Kahn and J. R. Barry, "Wireless infrared communications," *Proceedings of the IEEE*, vol. 85, no. 2, pp. 265–298, 1997.

[7] J. R. Barry, *Wireless Infrared Communications*. Springer US, Boston, MA, 1994.

[8] J. G. Proakis and M. Salehi, *Digital Communications*, 5th ed. McGraw-Hill, New York, 2008.

[9] K. Wang, A. Nirmalathas, C. Lim, and E. Skafidas, "Impact of background light induced shot noise in high-speed full-duplex indoor optical wireless communication systems," *Optics Express*, vol. 19, no. 22, pp. 21321–21332, 2011.

[10] S. D. Personick, "Receiver design for optical fiber systems," *Proceedings of the IEEE*, vol. 65, no. 12, pp. 1670–1678, 1977.

[11] T. Muoi, "Receiver design for high-speed optical-fiber systems," *Journal of Lightwave Technology*, vol. 2, no. 3, pp. 243–267, 1984.

[12] S. D. Personick, "Receiver design for digital fiber optic communication systems, I," *Bell System Technical Journal*, vol. 52, no. 6, pp. 843–874, 1973.

[13] S. D. Personick, "Receiver design for digital fiber optic communication systems, II," *Bell System Technical Journal,* vol. 52, no. 6, pp. 875–886, 1973.

[14] B. Leskovar, "Optical receivers for wide band data transmission systems," *IEEE Transactions on Nuclear Science*, vol. 36, no. 1, pp. 787–793, 1989.

5

Bit-Error-Rate Distribution and Outage of Indoor Optical Wireless Communications Systems

5.1 Introduction

Bit-error-rate (BER) distribution and outage is a critical measure of performance for indoor Optical Wireless Communications (OWC) systems. It is impacted by source layout, room structure, receivers' locations, and other factors. In the communications process, sources continuously emit light pulses in Lambertian pattern. Due to the room structure, a transmitted pulse undergoes multiple consecutive reflections on the surfaces, until it arrives at the receiver or attenuates to a negligible energy. Since multiple light sources, in a particular layout, are commonly used in a room for providing sufficient illumination, the indoor OWC channel is a spatially diverse channel to a specific user in the room. Because of multipath effects such as inter-source-interference (ISI) caused by reflections, the channel causes temporal distortions to the impulse. As the distortion strongly relies on receiver locations, the system will exhibit significantly different performance (BER) when the receiver is placed at different locations. In this chapter, we will consider all possible locations of users in the room and apply the MIMO channel model to simulate BER distribution and outage.

5.2 Simulation Parameters

The simulation assumes that there are multiple light sources on the room ceiling in a specific layout format. All these sources broadcast the same information. The floor is divided into many elements, and each of the elements is a possible receiver location, representing a possible user. For a particular receiver location, we generate the impulse response from the sources to the receiver at that location. Then, a data simulation is run based on the obtained impulse response to calculate BER. The simulation is carried for Intensity Modulation/Direct Detection (IM/DD). We repeat the process at other locations and obtain the BER of all possible

Short-Range Optical Wireless: Theory and Applications, First Edition. Mohsen Kavehrad,
M. I. Sakib Chowdhury and Zhou Zhou.
© 2016 John Wiley & Sons, Ltd. Published 2016 by John Wiley & Sons, Ltd.

user locations. Based on the BER data collected, we calculate the percentage of the room area that does not meet the outage requirement as the outage probability.

In practice, for mathematical convenience, we change the order slightly by using the MIMO modeling method [1], as described in Chapter 3. We gather all the impulse responses at all locations at first in a matrix $\mathbf{H}(t)$ as step 1, run the data simulations together to get the BERs in a matrix form as step 2, and calculate the BER outage as step 3. The sequential change leads to the results we desire. The theories of the MIMO modeling method are presented in detail in Chapter 3. We list the channel parameters used in the simulation in Table 5.1.

The parameters are chosen for the following reasons. For convenience, we set the transmission power of each light source to 680 lm, which is equal to 1 W at 555 nm. Receiver FOV is set to a large value as 60° to enlarge the receiving area as well as reduce blocking probability. Light emitting diode (LED) half power angle is set to 60° as the Lambertian mode number is equal to 1, which is the common value of most commercial LEDs. Reflection coefficients for the ceiling, walls, and floor are taken from previous research results. The room dimension is selected as that of an ordinary room.

Suppose there are n possible receiver locations. After the modeling, the impulse response matrix is obtained as $\mathbf{H}_{n\times1}(t)$. $\mathbf{H}_{n\times1}(t)$ is an $n\times1$ matrix, for which each entry represents the impulse response from the sources to one location.

Other parameters related to the BER simulation such as bit rate are shown in Table 5.2. Though it is higher than the maximum modulation rate of most available commercial LEDs, the computer simulation bit rate is set to 600 Mbps for two reasons. First, the focus in this simulation is on the channel and it is reasonable to idealize the source; second, by modern coding and equalization approaches, the available transmission rate of commercial LEDs approaches the value that is used here.

The layouts of 4, 6, and 9 sources are typical. $d[m]$ is the test data sequence. $s(t)$ is the transmission waveform, which is generated by OOK modulation of $d[m]$. $\mathbf{r}_{n\times1}(t)$ is the receiving waveform vector, while each entry is the received waveform at a specific location. Each entry of $\mathbf{r}_{n\times1}(t)$ is calculated from the convolution of $s(t)$ and corresponding entry of $\mathbf{H}_{n\times1}(t)$ plus noise. It can be presented as

$$\mathbf{r}_{n\times1}(t) = \mathbf{H}_{n\times1}(t) \otimes s(t) + n(t) \tag{5.1}$$

where $\otimes$ means convolution and $n(t)$ represents the noise.

Table 5.1   Wireless channel model parameters

Transmission power	680 lm
FOV	60°
LED half power angle	60°
Reflection coefficients (ceiling, wall, floor)	0.9, 0.5, 0.1
Room dimensions	6 m × 6 m × 3 m
Reflecting element size	0.1 m × 0.1 m
Time resolution	0.33 ns

Table 5.2   BER simulation parameters

Bit rate	600 Mbps
Noise	−105 dB mW

After recovering data $\hat{\mathbf{d}}_{n\times 1}[m]$ from $\mathbf{r}_{n\times 1}(t)$ by sampling and decision making, and comparing it with $\mathbf{d}_{n\times 1}[m]$, we obtain error vector $\mathbf{e}_{n\times 1}[m]$ as

$$\mathbf{e}_{n\times 1}[m] = \text{XOR}\left(\hat{\mathbf{d}}_{n\times 1}[m], \mathbf{d}_{n\times 1}[m]\right) \tag{5.2}$$

where XOR stands for exclusive-OR to each pair of entries. Each entry of $\mathbf{e}_{n\times 1}[m]$ is a "1-0" sequence where a "1" represents an error. Counting the number of 1s in each entry of $\mathbf{e}_{n\times 1}[m]$ and dividing it by the length of the corresponding entry, we obtain the BER vector $\mathbf{B}_{n\times 1}$. Each entry of $\mathbf{B}_{n\times 1}$ is the BER of the corresponding location. We set an acceptance threshold of BER for indoor wireless communications, for example, 10^{-5}. When an entry of $\mathbf{B}_{n\times 1}$ is greater than the threshold, it means that the corresponding location is in the outage state. The BER outage probability of the room is calculated by

$$\text{Outage} = \frac{\text{number of entries of}\left(\mathbf{B}_{n\times 1} > \text{threshold}\right)}{n} \tag{5.3}$$

5.3   Optimal Detection and BER Outage Analysis

In this section, an analytical method is used to predict visible light communications (VLC) BER distribution and outage caused by inter-source interference under arbitrary source layout. As diffuse reflections and overlapping distribution of light from different source layouts increase BER in certain scenarios while decreasing BER in other scenarios, we have applied the deterministic MIMO modeling method as described in Chapter 3 to calculate impulse response from the sources to the specific receiver location and then test data is used to compute the BER at that location. The process is repeated for all locations to obtain BER distribution in a room.

As mentioned in Chapter 3, the simulation method has a high computational complexity and requires a large volume of processing resources, especially for a high-resolution room model. That is because the method considers the Lambertian transmission between any pair of reflectors in each tier of diffusion; when room resolution increases by n times, the computational complexity increases by $n^2 \times n^2 = n^4$ times [2]. The test data may also consume a long time to process when high accurate BER is needed. The analytical method proposed in this chapter is based on Lambertian transmission, geometrical computation, and optimal detection theory.

5.3.1   Optimal Detection

As Lambertian transmission is described in Chapter 3 and the geometrical computation process is shown in the section, we provide a brief introduction of optimal detection here [3]. On-Off Keying (OOK) is the most popular VLC modulation scheme, which is basically binary amplitude modulation; therefore, we take binary antipodal signaling as an example to explain optimal detection.

In a binary antipodal signaling scheme $s_1(t) = s(t)$ and $s_2(t) = -s(t)$. The probabilities of messages 1 and 2 are p and $1-p$, respectively. The vector representations of the two signals are scaled as $s_1(t) = \sqrt{\varepsilon_s}$ and $s_2(t) = -\sqrt{\varepsilon_s}$, where ε_s is the energy in each signal. The decision region of signal 1, D_1, is given as

$$
\begin{aligned}
D_1 &= \left\{ r : r\sqrt{\varepsilon_s} + \frac{N_0}{2}\ln p - \frac{1}{2}\varepsilon_s > -r\sqrt{\varepsilon_s} + \frac{N_0}{2}\ln(1-p) - \frac{1}{2}\varepsilon_s \right\} \\
&= \left\{ r : r > \frac{N_0}{4\sqrt{\varepsilon_s}}\ln\frac{1-p}{p} \right\} \\
&= \left\{ r : r > r_{th} \right\}
\end{aligned}
\tag{5.4}
$$

where $N_0/2$ is the variance of the Gaussian noise and r_{th} is the threshold, defined as

$$
r_{th} = \frac{N_0}{4\sqrt{\varepsilon_s}}\ln\frac{1-p}{p}
\tag{5.5}
$$

Decision region of signal 2, D_2, is defined in a similar way. To derive the error probability for this system, we have

$$
\begin{aligned}
P_e &= \sum_{m=1}^{2} P_m \sum_{\substack{1 \le m' \le 2 \\ m' \ne m}} \int_{D_{m'}} p(r \mid s_m)\, dr \\
&= p\int_{D_2} p\!\left(r \mid s = \sqrt{\varepsilon_s}\right) dr + (1-p)\int_{D_1} p\!\left(r \mid s = -\sqrt{\varepsilon_s}\right) dr \\
&= p\int_{-\infty}^{r_{th}} p\!\left(r \mid s = \sqrt{\varepsilon_s}\right) dr + (1-p)\int_{r_{th}}^{\infty} p\!\left(r \mid s = -\sqrt{\varepsilon_s}\right) dr \\
&= pP\!\left[N\!\left(\sqrt{\varepsilon_s}, \frac{N_0}{2}\right) < r_{th} \right] + (1-p)P\!\left[N\!\left(-\sqrt{\varepsilon_s}, \frac{N_0}{2}\right) > r_{th} \right] \\
&= pQ\!\left(\frac{\sqrt{\varepsilon_s} - r_{th}}{\sqrt{N_0/2}} \right) + (1-p)Q\!\left(\frac{r_{th} + \sqrt{\varepsilon_s}}{\sqrt{N_0/2}} \right)
\end{aligned}
\tag{5.6}
$$

In the special case where $p = (1/2)$, we have $r_{th} = 0$, and the error probability simplifies to

$$
P_e = Q\!\left(\sqrt{\frac{2\varepsilon_s}{N_0}} \right)
\tag{5.7}
$$

Supposing that there is an interference signal at the next symbol period with energy ε_i, the error probability becomes

$$
P_e = \frac{1}{2}Q\!\left(\sqrt{\frac{2}{N_0}}\left(\sqrt{\varepsilon_s} - \sqrt{\varepsilon_i}\right) \right)
\tag{5.8}
$$

where the coefficient 1/2 indicates that the interference symbol is different from the transmitted symbol by a probability of 1/2.

5.3.2 BER Analysis

As described earlier, there are two kinds of impulse distortions in VLC systems: multipath effects and inter-source interference. In most room areas, inter-source impulse distortion dominates. It can be modeled by multiple LOS Lambertian transmissions from different sources. The inter-source impulse distortion model can be further used to determine room BER distribution.

Suppose there are N sources $S_1, S_2, \ldots, S_N$ on the ceiling, streaming data to receiver R. The coordinate of source i is $(s_{ix}, s_{iy}, s_{iz})(i = 1, 2, \ldots, N)$ and the coordinate of the receiver is (r_x, r_y, r_z). We define the distance matrix $\mathbf{H}_D$ and the power matrix $\mathbf{H}_P$ as,

$$\mathbf{H}_D = [d_1, d_2, \ldots, d_N] \tag{5.9}$$

$$\mathbf{H}_P = [p_1, p_2, \ldots, p_N] \tag{5.10}$$

where d_i and p_i indicate the distance and the received power from source $S_i (i = 1, 2, \ldots, N)$ to the receiver R, respectively. Applying the geometric method, it is straightforward to obtain d_i as

$$d_i = \left\| (s_{ix}, s_{iy}, s_{iz}) - (r_x, r_y, r_z) \right\|, \quad (i = 1, 2, \ldots, N) \tag{5.11}$$

Using the LOS Lambertian model, we calculate p_i as

$$
\begin{aligned}
p_i &= \frac{n+1}{2\pi} \cos^n(\varphi_i) \cos(\theta_i) A_{Ri} / d_i^2 \; \mathrm{rect}\left(\frac{\theta_i}{\mathrm{FOV}}\right) \\
&= \frac{n+1}{2\pi} \cos^{n+1}(\varphi_i) A_{Ri} / d_i^2 \; \mathrm{rect}\left(\frac{\theta_i}{\mathrm{FOV}}\right) \\
&= \frac{n+1}{2\pi} \left(\frac{s_{zi} - r_z}{d_i}\right)^{n+1} A_{Ri} / d_i^2 \; \mathrm{rect}\left(\frac{\theta_i}{\mathrm{FOV}}\right) \\
&= \frac{n+1}{2\pi} A_{Ri} \left(s_{iz} - r_2\right)^{n+1} d_i^{-3-n} \; \mathrm{rect}\left(\frac{\theta_i}{\mathrm{FOV}}\right)
\end{aligned}
\tag{5.12}
$$

We consider LOS transmission from source $S_i (i = 1, 2, \ldots, N)$ to receiver R; since the ceiling plane is parallel to the floor plane, we have receiver incident angle θ_i equal to source transmission angle φ_i. Then, $\cos(\varphi_i)$ can be calculated by trigonometric relations as

$$\cos(\varphi_i) = \frac{s_{zi} - r_z}{d_i} \tag{5.13}$$

In this chapter, we assume Lambertian pattern order $n = 1$, incident angle θ_i is always smaller than receiver FOV, and all receivers have the same area A_R. As a result, we have

$$p_i = \frac{1}{\pi} \frac{A_R \left(s_{iz} - r_z \right)^2}{d_i^4} \tag{5.14}$$

From $\mathbf{H}_D$ and $\mathbf{H}_p$, we know that the impulse response $h(t)$ from the sources $S_1, S_2, \ldots, S_N$ to receiver R consists of N peaks, with different amplitudes arriving at different times as

$$h(t) = \sum_{i=1}^{N} \frac{1}{\pi} \frac{A_R \left(s_{iz} - r_z \right)^2}{d_i^4} \delta \left(t - \frac{d_i}{c} \right) \tag{5.15}$$

Let the transmitted signal be $s(t)$; then the received signal is given by

$$\begin{aligned} r(t) &= h(t) \otimes s(t) + n(t) \\ &= \frac{A_R}{\pi} \sum_{i=1}^{N} \frac{\left(s_{iz} - r_z \right)^2}{d_i^4} s \left(t - \frac{d_i}{c} \right) + n(t) \end{aligned} \tag{5.16}$$

We define the earliest arrived light impulse as the user impulse and suppose it comes from source S_0. It satisfies $S_0 = \arg\min_{1 \le i \le N} \left(d_i \right)$.

When the peaks arrive after the user peak no later than a symbol period T, as $s(t)$ remains unchanged, these peaks will enforce the user peak. Therefore, the received user energy is

$$\varepsilon_s = \int_0^T \left(\frac{A_R}{\pi} \sum_{\substack{i \\ 0 < d_i - d_0 \le cT}} \frac{\left(s_{iz} - r_z \right)^2}{d_i^4} s \left(t - \frac{d_i}{c} \right) \right)^2 dt \tag{5.17}$$

We consider the peaks arriving after the user peak between T and $2T$, and ignore the later peaks, since they are significantly weaker. The interference optical energy is

$$\varepsilon_i = \int_T^{2T} \left(\frac{A_R}{\pi} \sum_{\substack{i \\ cT < d_i - d_0 \le 2cT}} \frac{\left(s_{iz} - r_z \right)^2}{d_i^4} s \left(t - \frac{d_i}{c} \right) \right)^2 dt \tag{5.18}$$

Since by most optical detection approaches, optical energy will be converted into current for processing, optical energy is proportional to the square root of signal energy; therefore, we square each term of optical energy to calculate signal energy. Suppose binary antipodal modulation is used. From (5.8), the BER at that receiver location is

$$P_e = \frac{1}{2} Q\left(\sqrt{\frac{2}{N_0}} \left(\sqrt{\varepsilon_s} - \sqrt{\varepsilon_i} \right) \right)$$

$$= \frac{1}{2} Q\left(\frac{A_R}{\pi} \sqrt{\frac{2}{N_0}} \left(\sqrt{ \int_0^T \left(\frac{A_R}{\pi} \sum_{\substack{i \\ 0 < d_i - d_0 \le cT}} \frac{(s_{iz} - r_z)^2}{d_i^4} s\left(t - \frac{d_i}{c} \right) \right)^2 dt } \right. \right.$$
$$\left. \left. - \sqrt{ \int_T^{2T} \left(\frac{A_R}{\pi} \sum_{\substack{i \\ cT < d_i - d_0 \le 2cT}} \frac{(s_{iz} - r_z)^2}{d_i^4} s\left(t - \frac{d_i}{c} \right) \right)^2 dt } \right) \right) \qquad (5.19)$$

The coefficient 1/2 means the interference symbol is different from user symbol by a probability of 50%. Repeating the process for all locations, we can calculate the BER distribution and outage.

5.4   Simulation Results (Receiver FOV $= 60°$)

When a light pulse travels in an indoor environment, it experiences reflections. When the number of these reflections is large, multipath effects become significant and the pulse spreads in time domain. If multiple sources are applied, inter-source interference occurs. The interfering sources produce additional delayed pulses to the ideal impulse response. We conduct our simulations through the following steps:

(a) Locate l sources on the ceiling in a specific layout for spatial diversity.
(b) Generate indoor VLC spatial diversity matrix $\mathbf{H}_{n \times 1}(t)$.
(c) The receiver plane is uniformly divided into n elements as receiver locations. Each row of $\mathbf{H}_{n \times 1}(t)$ refers to the impulse response from sources to a certain receiver.
(d) Generate test data $d[m]$ and corresponding waveform $s(t)$ by OOK modulation.
(e) Calculate received waveform $\mathbf{r}_{n \times 1}(t)$ by $\mathbf{r}_{n \times 1}(t) = \mathbf{H}_{n \times 1}(t) \otimes s(t) + n(t)$. $\otimes$ denotes convolution and $n(t)$ is Gaussian noise.
(f) Recover data $\hat{\mathbf{d}}_{n \times 1}[m]$ from $\mathbf{r}_{n \times 1}(t)$ by sampling and decision making, where each element of $\hat{\mathbf{d}}_{n \times 1}(m)$ is the received data at each receiver location.
(g) Compare each element of $\hat{\mathbf{d}}_{n \times 1}[m]$ with $\mathbf{d}_{n \times 1}[m]$ to obtain error vector $\mathbf{e}_{n \times 1}[m]$ by using (5.2).
(h) Each entry of $\mathbf{e}_{n \times 1}[m]$ is a sequence of 1s and 0s where a "1" represents an error. Counting the number of 1s in each entry of $\mathbf{e}_{n \times 1}[m]$, we estimate the BER vector $\mathbf{B}_{n \times 1}$. Each entry of $\mathbf{B}_{n \times 1}$ is the BER of the corresponding location.
(i) We set an acceptance threshold of BER for indoor wireless communications, for example, 10^{-5}. When an entry of $\mathbf{B}_{n \times 1}$ is greater than the threshold, it means that the corresponding location is in an outage state. We compute outage probability by (5.3).

The parameters used are given in Tables 5.1, 5.2 and 5.3.

Table 5.3 Source locations used in BER simulation

Locations of four sources	(1.5 m, 1.5 m, 3 m), (1.5 m, 4.5 m, 3 m), (4.5 m, 1.5 m, 3 m), (4.5 m, 4.5 m, 3 m)
Locations of six sources	(1 m, 1.5 m, 3 m), (3 m, 1.5 m, 3 m), (5 m, 1.5 m, 3 m), (1 m, 4.5 m, 3 m), (3 m, 4.5 m, 3 m), (5 m, 4.5 m, 3 m)
Locations of nine sources	(1 m, 1 m, 3 m), (3 m, 1 m, 3 m), (5 m, 1 m, 3 m), (1 m, 3 m, 3 m), (3 m, 3 m, 3 m), (5 m, 3 m, 3 m), (1 m, 5 m, 3 m), (3 m, 5 m, 3 m), (5 m, 5 m, 3 m)

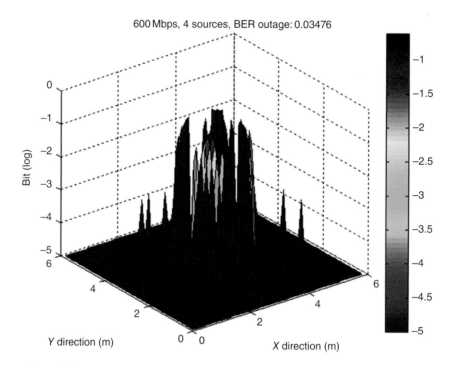

Figure 5.1 BER distribution and outage of a layout having four sources, FOV = 60°.

5.4.1 BER Distribution and Outage

Figures 5.1, 5.2, 5.3, 5.4, 5.5, and 5.6 show the BER distribution and outage of layouts having four sources, six sources, and nine sources, respectively. In these figures, X and Y values indicate the position of a receiver on the floor; Z value is the corresponding logarithmic BER. The higher the BER is in a location, the worse communication quality is expected for the user located there. X–Y views of the figures clearly demonstrate the BER distribution on the floor. Setting a BER threshold, for instance, 10^{-5} as we do, we can easily calculate the BER outage for the room in a specific source layout.

Comparing Figures 5.1, 5.2, 5.3, 5.4, 5.5, and 5.6, we find that high BER exists in the overlapping areas of the sources and these areas enlarge as the number of sources increases. The results can be explained as follows. When the number of sources increases, the layout of the sources becomes denser and there are more illumination overlapping areas on the floor. In the

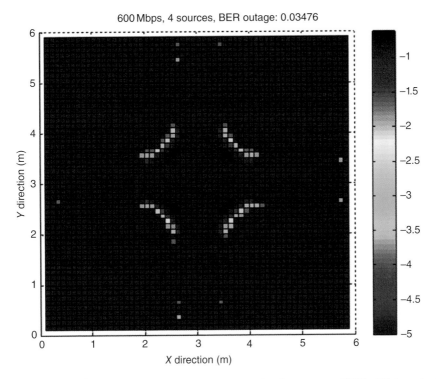

Figure 5.2 BER distribution and outage of a layout having four sources, FOV = 60° (top view).

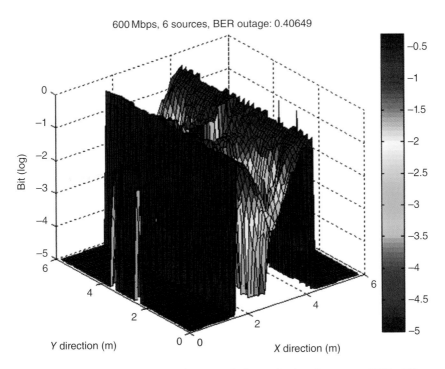

Figure 5.3 BER distribution and outage of a layout having six sources, FOV = 60°.

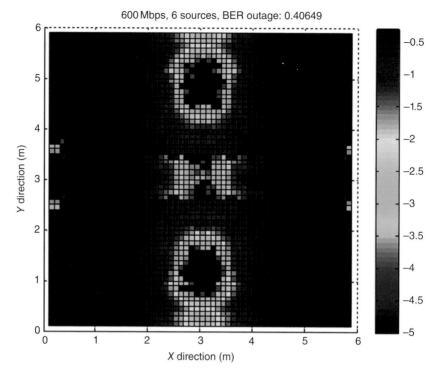

Figure 5.4 BER distribution and outage of a layout having six sources, FOV = 60° (top view).

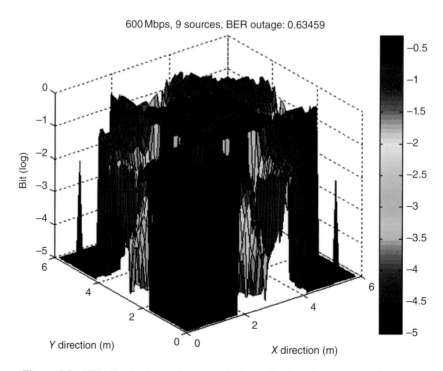

Figure 5.5 BER distribution and outage of a layout having nine sources, FOV = 60°.

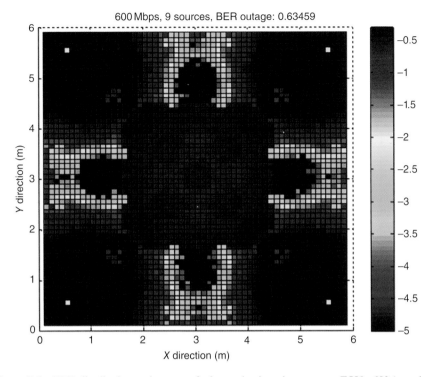

Figure 5.6   BER distribution and outage of a layout having nine sources, FOV = 60° (top view).

overlapping areas, communication is subject to inter-source interference, and the impulse response from sources at these locations will be distorted more significantly. The distortion of impulse response will cause ISI, and thus a high BER. The simulation demonstrates that dense source layout will increase BER outage for a VLC system.

Another interesting observation is that there are less high BER regions near room corners. This holds as layout density increases. The result shows that the interference from multipath effect is insignificant on BER outage. Though multipath impulse response degrades with the increasing number of sources, the BER outage is not significantly impacted by multipath impulse response spread.

5.4.2   Impulse Response Distortion

By computer simulations, we may work out the total impulse response from the source to any possible receiver location. In most locations, the impulse response has little time spread. However, we still find that there are two kinds of locations exhibiting considerable impulse distortion, as shown in Figure 5.7: the first location is the room corner (location A (0.2 m, 0.1 m)) and the second location is the overlapping area of light footprints (location B (2.2 m, 2.4 m)).

At the room corner, the spread of impulse response is the result of multipath effects of light reflections. In the corner area, the receiver captures reflected lights, which experiences different reflection paths, and causes the longer tail in the impulse response. What is worth mentioning is that the tail has much lower power compared with the peak. It can be shown that this kind of distortion exhibits less significant influence on VLC.

(a)

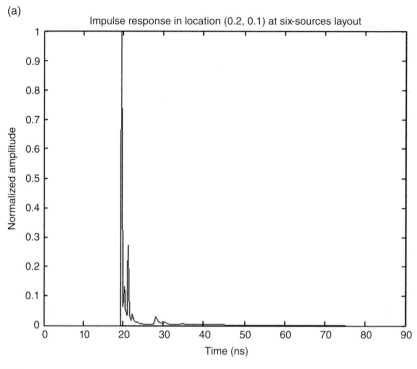

(b)

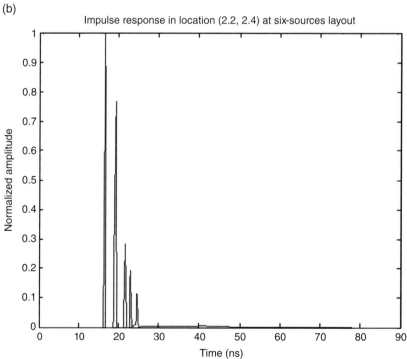

Figure 5.7 Impulse response distortion by (a) multipath effect (b) inter-source interference (FOV = 60°).

In the overlapping areas, we can see two high sharp peaks in the impulse response. The reason for the multiple peaks is that light from different sources reaches the receiver via different LOS paths, and thus the time difference of the arrivals causes the multiple peaks. This is an important influencing factor for VLC.

5.5   Simulation Results (Receiver FOV = 30°)

To further explore the impact of inter-source interference to BER distribution, we repeated the simulation with receiver FOV = 30°. The results are given in Figures 5.8, 5.9, 5.10, 5.11, 5.12, and 5.13. We find that the BER outage reduces from 0.4065 to 0.1709 in six-sources layout and from 0.6346 to 0.3453 in nine-sources layout, respectively. That is because in many locations, small receiver FOV reduces interference power and BER; therefore, BER outage decreases. This effect can be clearly observed by comparing Figure 5.7 with Figure 5.14. They are impulse responses at same locations, with different FOVs. In Figure 5.14, most of the inter-source interference peaks are filtered out. An interesting finding is that in four-sources scenario, BER outage of FOV = 30° is higher than that of FOV = 60°. That is because a small FOV also decreases total received power. When the number of sources is small and a small FOV is applied, the BER may increase due to insufficient signal-to-noise ratio (SNR), in certain locations. This point should be paid attention to, when we want to reduce inter-source interference by utilizing small FOV receivers.

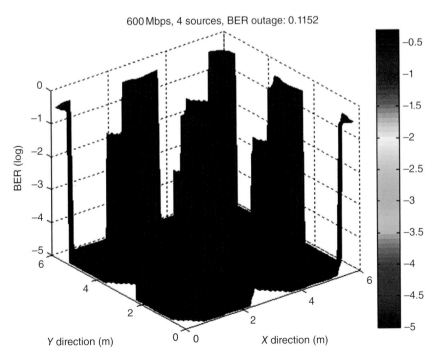

Figure 5.8   BER distribution and outage of a layout having four sources, FOV = 30°.

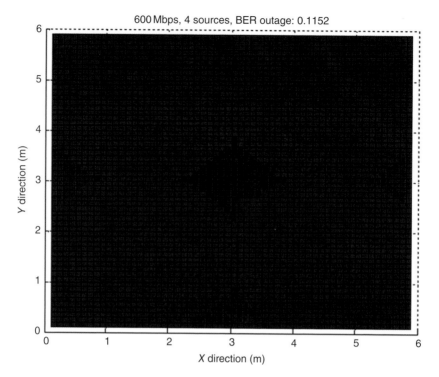

Figure 5.9 BER distribution and outage of a layout having four sources, FOV = 30° (top view).

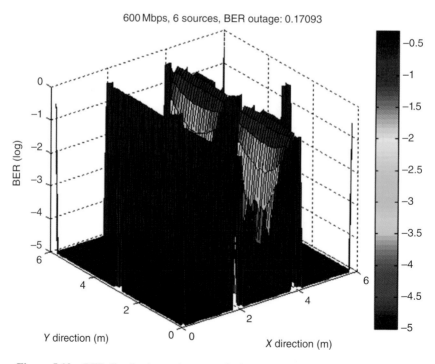

Figure 5.10 BER distribution and outage of a layout having six sources, FOV = 30°.

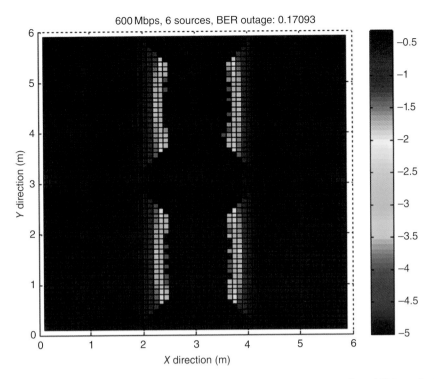

Figure 5.11 BER distribution and outage of a layout having six sources, FOV = 30° (top view).

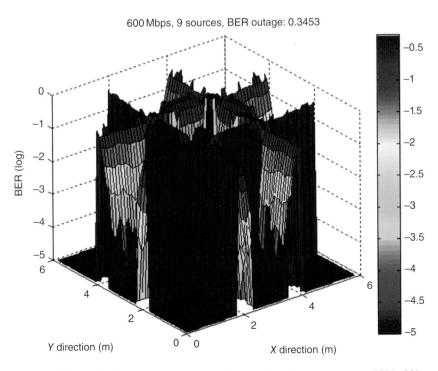

Figure 5.12 BER distribution and outage of a layout having nine sources, FOV = 30°.

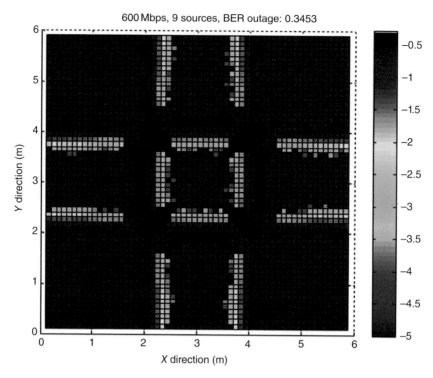

Figure 5.13   BER distribution and outage of a layout having nine sources, FOV = 30° (top view).

5.6   Analytical Results and Comparisons

As shown in the previous section, BER caused by inter-source interference can be found by (5.17–5.19). By this method, we can analytically calculate the BER distribution and outage. The results of four-sources and six-sources of FOV = 60° are shown in Figures 5.15, 5.16, 5.17, and 5.18. Comparing with Figures 5.1, 5.2, 5.3, and 5.4, we can see a good match in BER distribution and outage.

5.7   Conclusions

In this chapter, we have investigated the feasibility of VLC systems. This technology is a high-speed, energy-efficient, and secure solution to RF band congestion. In a general home environment, the major impact factors to VLC are multipath effects and inter-source interference, which degrade communications performance by causing impulse response distortion. To further investigate these factors, at first, we established an indoor optical wireless model by tracking light pulses experiencing reflections. After the channel model was obtained, we used data simulation to statistically calculate BER distribution and outage. By observing the results, we found that multipath effect exists at room corner locations and inter-source interference exists at the overlapping area of light footprints. Moreover, the influence of inter-source interference is more significant than multipath effects.

(a)

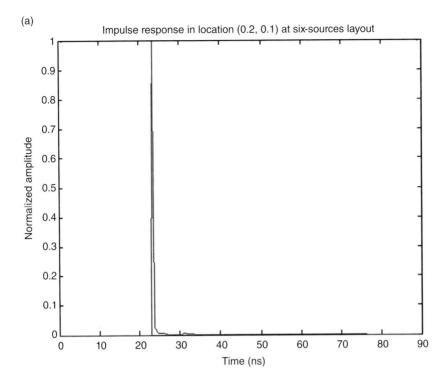

(b)

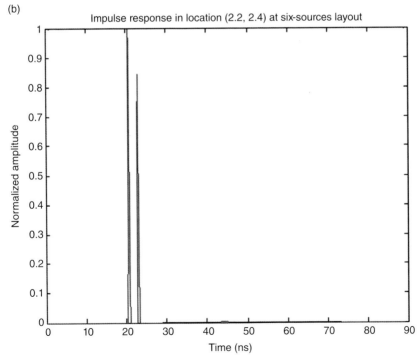

Figure 5.14   Impulse response distortion by (a) multipath effect and (b) inter-source interference (FOV = 30°).

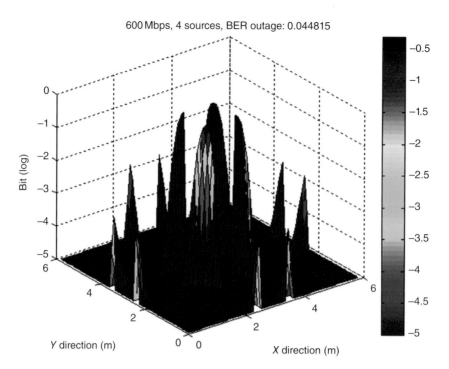

Figure 5.15 BER distribution and outage of a layout having four sources by analytical method, FOV = 60°.

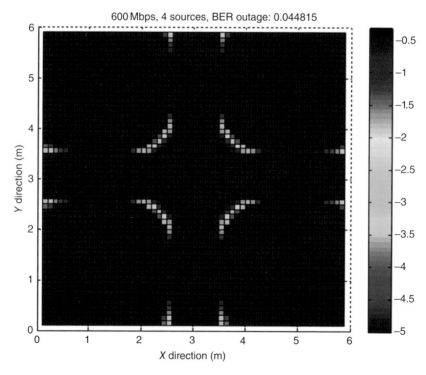

Figure 5.16 BER distribution and outage of a layout having four sources by analytical method, FOV = 60° (top view).

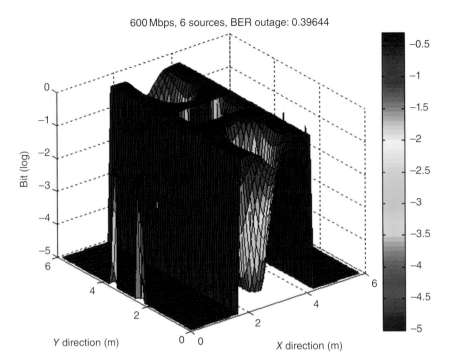

Figure 5.17 BER distribution and outage of a layout having six sources by analytical method, FOV = 60°.

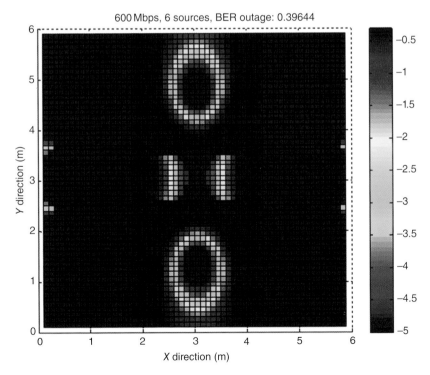

Figure 5.18 BER distribution and outage of a layout having six sources by analytical method, FOV = 60° (top view).

References

[1] Y. A. Alqudah and M. Kavehrad, "MIMO characterization of indoor wireless optical link using a diffuse-transmission configuration," *IEEE Transactions on Communications*, vol. 51, no. 9, pp. 1554–1560, 2003.

[2] J. R. Barry, J. M. Kahn, W. J. Krause, E. A. Lee, and D. G. Messerschmitt, "Simulation of multipath impulse response for indoor wireless optical channels," *IEEE Journal on Selected Areas in Communications*, vol. 11, no. 3, pp. 367–379, 1993.

[3] J. G. Proakis and M. Salehi, *Digital Communications*, 5th ed. McGraw-Hill, Boston, MA, 2008, p. 1150.

6

Orthogonal Frequency-Division Multiplexing (OFDM) for Indoor Optical Wireless Communications

6.1 Introduction

Orthogonal frequency division multiplexing (OFDM) is currently being used predominantly in radio frequency (RF) mobile broadband communication systems because of its ability to combat intersymbol interference (ISI) and robustness against frequency-selective fading caused by multipath wireless channel. Wireless mobile standards such as 3G and 4G long-term evolution (LTE) use orthogonal frequency division multiple access (OFDMA) as a multiplexing/modulation scheme. Despite its many advantages such as single-tap frequency domain equalization and fast discrete time implementation, OFDM suffers from certain disadvantages such as high peak-to-average power ratio (PAPR) and high sensitivity to carrier frequency offset (CFO). Although OFDM has solved problems such as multipath fading, it cannot solve the emerging problems such as scarcity of RF spectrum for mobile wireless broadband applications. Optical wireless communications (OWC) can go a long way to solve the problems of RF spectrum scarcity, and hence OFDM is also being considered as a candidate for OWC systems, especially those based on visible light communications (VLC) as it offers robustness against multipath caused by diffuse indoor optical wireless (OW) channel. As we have discussed earlier, one way to realize VLC is intensity modulation direct detection (IM/DD).

Although the major difference between RF and OW-based OFDM lies in the front end of transmitter and receiver, due to the unipolar nature of optical intensity in the IM/DD system, methods of generating baseband OFDM signal, techniques to reduce PAPR and timing synchronization schemes for RF cannot be directly applied to optical OFDM systems and therefore must be revisited.

Short-Range Optical Wireless: Theory and Applications, First Edition. Mohsen Kavehrad, M. I. Sakib Chowdhury and Zhou Zhou.
© 2016 John Wiley & Sons, Ltd. Published 2016 by John Wiley & Sons, Ltd.

In this chapter, we develop some techniques to reduce high PAPR in OFDM-based OW systems as the nonlinear characteristics of LED transmitters can severely affect system performance. We look into various precoding-based PAPR reduction techniques. We then analyze the performance of various OFDM-based OW schemes in multipath diffuse indoor wireless channel. We also compare the performance of conventional schemes with a precoded version.

6.2 OFDM Overview

6.2.1 Basic OFDM System

The basic idea behind OFDM is to transmit a serial stream of data on N multiple parallel channels of narrow bandwidth [1]. This is in contrast to the conventional serial data transmission system where each symbol occupies the entire available bandwidth and is transmitted for T_S symbol period. Thus in OFDM, each data symbol is transmitted for a longer duration $T_B = NT_S$ where T_B is block period.

By transmitting data in parallel, we can alleviate a number of problems that we face in serial data transmission systems. In parallel transmission, each stream occupies a small portion of the available bandwidth. Usually the bandwidth is divided into N nonoverlapping subchannels. To obtain more spectral efficiency, the sub-channels are allowed to overlap with an orthogonality constraint so that data modulated on individual channels can be easily recovered at the receiver.

Parallel transmission causes a fade to spread over many symbols that are not adjacent. Thus, a burst error caused by Rayleigh fading is randomized over several symbols improving the bit error performance of the system. The main advantage of the OFDM parallel transmission is that each symbol is transmitted for a longer duration, which makes the transmission less sensitive to delay spread.

6.2.2 System Operation

A simple continuous time OFDM communication system block diagram is shown in Figure 6.1. Serial data stream is input to the encoder that produces the complex symbols d_i according to the modulation scheme used. The complex data symbol can be represented by

$$d_i = a_i + jb_i \tag{6.1}$$

where a_i and b_i are real values that represent the in-phase and quadrature components, respectively. In the conventional serial data transmission system the transmitted signal would be represented by

$$D(t) = \sum_i \left[a_i \cos(\omega_c t) + b_i \sin(\omega_c t) \right] g(t - iT_S) \tag{6.2}$$

In OFDM the baseband data waveform is represented by

$$s(t) = \sum_i \left[\sum_{k=0}^{N-1} \left\{ d_{i,k} e^{j2\pi f_k t} \right\} \right] g(t - iT_B) \tag{6.3}$$

(a)

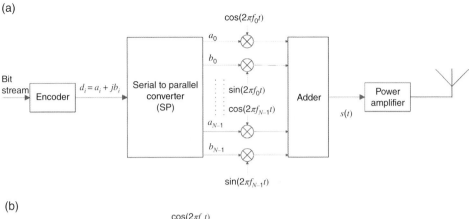

(b)

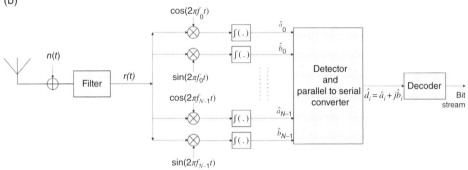

Figure 6.1   A simple continuous time OFDM (a) transmitter and (b) receiver.

where $g(t)$ is a pulse, usually rectangular in shape given by

$$g(t) = \begin{cases} 1, & 0 \le t \le T_B \\ 0, & \text{elsewhere} \end{cases} \tag{6.4}$$

where $f_k = k/T_B$ is the frequency of the k-th subcarrier from the set of subcarriers $\{e^{j2\pi f_k t}, k = 0,1,\ldots,N-1\}$ and N data symbols are transmitted in parallel during the i-th block. The subcarrier spacing is chosen as $\Delta f = 1/T_B$ Hz. This spacing makes the adjacent subcarriers overlap while satisfying the orthogonality condition, which makes the recovery/demodulation of each subcarrier easier at the receiver.

The frequency domain representation of one block of OFDM data can be obtained using the Fourier Transform of 0th block,

$$s(t) = \sum_{k=0}^{N-1} \left\{ d_{0,k} e^{j2\pi f_k t} \right\} g(t), \quad \text{for } i = 0$$

$$S(f) = F\{s(t)\}$$

$$= F\left\{ \sum_{k=0}^{N-1} \left\{ d_{0,k} e^{j2\pi f_k t} \right\} g(t) \right\} \tag{6.5}$$

$$= e^{j2\pi f T_B/2} \sum_{k=0}^{N-1} d_{0,k} \text{sinc}\left(\left[f - k/T_B \right] T_B \right)$$

The expression shows that in frequency domain the subcarriers will be tightly packed and overlapping but will not be interfering at the $f_k = k / T_B$ spacing where one subcarrier will have peak while all others will be zero. Thus we see that OFDM transmits N data symbols in parallel using multiple carrier frequencies with narrow bandwidths.

6.2.3 Discrete Time Implementation of OFDM

To implement an OFDM system in continuous time, we need multiple modulators and filters that increase the equipment complexity. Multiple banks of correlators required at the receiver make it very difficult to be realized practically. However, a great amount of equipment reduction can be obtained by implementing OFDM modulation using IFFT. It can be seen mathematically that a baseband OFDM waveform is in fact IFFT of original waveform followed by a D/A conversion. Mathematically,

$$s(t) = \sum_{k=0}^{N-1} d_{0,k} e^{j2\pi f_k t}$$

Sampling it at $t = mT_B / N$,

$$s(t) = \sum_{k=0}^{N-1} d_{0,k} e^{2\pi f_k mT_B / N}$$

$$y(m) = s(t)\Big|_{t=mT_B/N} = \sum_{k=0}^{N-1} d_{0,k} e^{2\pi km / N} \tag{6.6}$$

where we see that the sequence $y(m)$ is effectively the IFFT of the data vector $d_{i,k}$. When the sequence $y(m)$ is passed through a digital-to-analog (D/A) converter, we get the same waveform $s(t)$. At the receiver side, reverse operation is performed by first sampling the waveform $s(t)$ and then taking FFT of the samples, giving us the complex symbol estimates $\hat{d}_{i,k}$ which will be used to generate the serial bit stream that was originally transmitted. Mathematically,

$$d_{0,k} = \frac{1}{N} \sum_{m=0}^{N-1} y(m) e^{-j2\pi km / N}, \quad k = 0,1,\ldots,N-1 \tag{6.7}$$

Both FFT and IFFT can be implemented using computationally efficient computer algorithms. Thus, a great amount of simplification is achieved by using these techniques as compared to performing modulation/demodulation in continuous time using N oscillators. A system block diagram for discrete time implementation is given in Figure 6.2.

6.2.4 Drawbacks of OFDM

Although OFDM is being used in many RF applications [2] and is being considered as a candidate for high speed OW systems, it suffers from certain disadvantages as follows.

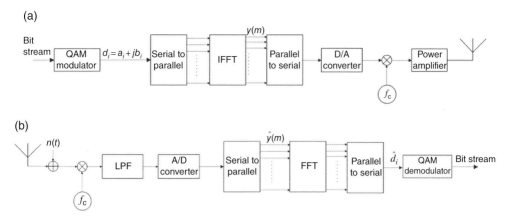

Figure 6.2   Discrete time implementation of OFDM (a) transmitter and (b) receiver.

6.2.4.1   High PAPR

As OFDM is a multicarrier technique, output signal has a very high PAPR, which requires a very wide dynamic range linear power amplifier (PA). Designing linear PAs with a wide dynamic range is very expensive. Therefore, the PAPR of the OFDM output signal has to be reduced in order to use nonlinear PA, which is power efficient and inexpensive. High PAPR is also a problem in OW communication which uses LEDs as transmitters. This is due to the fact that LED transfer characteristics are also nonlinear. Therefore, some strategies have to be used to reduce PAPR of the OFDM output signal for both RF and OW systems to design economical communication systems.

6.2.4.2   Sensitivity to Carrier Frequency Offset (CFO)

The second major drawback of OFDM is its high sensitivity to CFO. In OFDM, individual subcarriers are overlapping and orthogonal to each other. A slight difference in the carrier frequency or sampling rate at the receiver will disturb orthogonality among the subcarriers and will cause interference to neighboring subcarriers. This will reduce the signal-to-noise ratio (SNR) and will deteriorate system performance. CFO can occur in mobile receivers moving at very high speed. High speed causes signal frequency to increase or decrease, depending on the direction of motion. If the receiver is moving toward the transmitter, the frequency will increase and if the receiver is moving away from the transmitter, the frequency will decrease. In either scenario, CFO will occur as there will be a shift in the frequency of the received signal due to motion of the mobile user. CFO has to be countered in an effective way in order to receive the signal without interference. CFO can also occur due to shift in the frequency of the local oscillator (LO) at the receiver. This shift can be determined through training symbols and can be easily fixed. However, CFO cancellation in case of Doppler is not an easy task, especially at the base station where signal from multiple users is received and each user is moving with a different velocity. Estimating the Doppler shift for every user is very difficult. Therefore, some other strategies have to be investigated to cancel the CFO at the receiver. For OW systems, as the users are not mobile at high speeds, CFO problems due to Doppler are fortunately very small or nonexistent. However, shift in the frequency of the LO is still needed to be taken care of.

6.3   OFDM-Based OW Systems

OFDM is also being considered as a candidate for indoor OW systems, especially in IM/DD systems and has gained significant attention because of the multipath nature of indoor OW channel [3–6]. Multipath in an indoor environment causes overlapping of light signals and results in signal distortion [7, 8]. This severely degrades system performance.

In RF-based OFDM systems, output signal is bipolar and complex. This signal cannot be easily transmitted in an OW system as light intensity cannot be negative and we cannot transmit a complex signal using a single optical transmitter such as LED [9]. Therefore, output OFDM signal has to be made real and positive to make it suitable for optical transmission. Hermitian symmetric input data to OFDM block generates a real output signal. However, to make the signal positive, several OFDM schemes have been proposed for IM/DD OW systems. Among them, one is called DC-Biased OFDM [10] wherein we use a DC bias to make the output signal positive. Other schemes involve clipping the negative part of the output signal. PAM-DMT [11] is one of these clipping-based schemes where we modulate the complex part of each subcarrier with a real symbol, which will result in clipping noise to fall on the real part of the same subcarrier. Another clipping-based scheme known as asymmetrically clipped (AC) optical OFDM (ACO-OFDM) uses only odd subcarriers modulated by complex constellation symbols [12, 13]. This will result in clipping noise to fall only on even subcarriers. Therefore, in both clipping-based strategies, the clipping noise is always orthogonal to the transmitted symbols, which will enable easy recovery of the desired data at the receiver. Another technique called discrete Hartley transform (DHT)-based optical OFDM [14] uses real input symbols and generates a real bipolar output signal using Hartley transform. The characteristics of output signal are similar to those in ACO-OFDM.

In this chapter, we focus only on these three AC-based OFDM techniques. A generic block diagram of AC-based OFDM system is shown in Figure 6.3. Only constellation mapping, mapping and zero insertion, frequency to time transformation (FT), time to frequency (TF) domain transformation, and extract symbols blocks will perform different operations on the input data for each scheme. The rest of the transmitter and receiver blocks will remain the same.

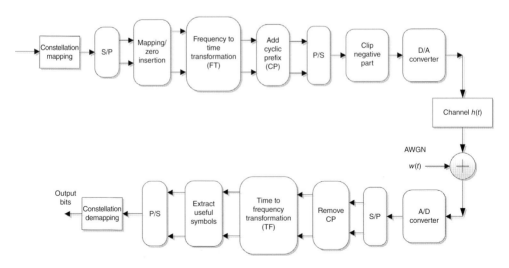

Figure 6.3   A generalized block diagram of asymmetric clipped-based OFDM systems.

6.3.1 ACO-OFDM

In this OFDM-based system, data are transmitted in the form of blocks having duration of T seconds. Each block consists of $M = N/4$ complex symbols drawn from a complex 2D constellation mapping scheme like 4-, 16-, or 64-QAM, which will modulate only odd subcarriers in the first half of N subcarriers. N is the total number of subcarriers available and is equal to the size of IFFT. In ACO-OFDM, the FT block will perform IFFT operations on input data. The conjugates of these symbols modulate the odd subcarriers of the second half of N subcarriers to meet the Hermitian symmetry requirements. Therefore, the input data vector to the IFFT block will look like $\mathbf{X} = \left[0, X_0, 0, X_1, 0, \ldots, X_{N/2-1}, 0, X_{N/2-1}^*, 0, \ldots, X_0^* \right]$, where $X_k = a_k + ib_k$ and a_k, b_k are real and imaginary parts of the complex symbol, respectively. The first (DC) and $N/2$-nd subcarriers are set to zero to obtain a real output signal. The time-domain output signal is generated by taking the IFFT of the input vector,

$$x_n = \frac{1}{N} \sum_{k=0}^{N-1} X_k e^{j 2\pi \frac{k}{N} n} \tag{6.8}$$

A Cyclic Prefix (CP) is added to this discrete time output signal. x_n is bipolar and anti-symmetric. We clip the negative part of this signal to generate a unipolar signal $\lfloor x_n \rfloor_c$ given by

$$\lfloor x_n \rfloor_c = \begin{cases} x_n & \text{if} \quad x_n \geq 0 \\ 0 & \text{if} \quad x_n < 0 \end{cases} \tag{6.9}$$

$\lfloor x_n \rfloor_c$ finally passes through a D/A converter to generate a continuous time-domain signal and ultimately modulates the intensity of the optical transmitter such as an LED. Clipping noise generated by clipping negative half of time domain signal falls only on even subcarriers. Therefore, the transmitted symbols are not affected by clipping noise, which enables easy recovery of transmitted data at the receiver.

At the receiver, an optical detector converts the intensity into an electrical signal $\tilde{x}(t)$. This signal gets corrupted by electronic noise generated by the electronic components and ambient noise from the surrounding light sources. This noise $w(t)$ is usually modeled as additive white Gaussian noise (AWGN). The noise-corrupted signal is then passed through an A/D converter to generate a discrete time signal $\tilde{x}_n$.

$$\tilde{x}_n = \lfloor x_n \rfloor_c + w_n \tag{6.10}$$

where w_n is discrete time version of AWGN. After removing CP, the TF block performs N-point FFT operation on the input discrete time samples. The noise corrupted constellation symbols are extracted from the FFT output and demapped to generate the output bits.

6.3.2 PAM-DMT

In this OFDM-based scheme, $N/2$ symbols drawn from a real mapping scheme like PAM are used to modulate the complex part of each subcarrier. However, the DC and $N/2$-nd subcarrier are not modulated to fulfill the Hermitian symmetry requirements. Therefore, the data vector forming the input to the FT block will be $\mathbf{Y} = \left[0, Y_0, Y_1, Y_2, \ldots, Y_{N/2-1}, 0, Y^*_{N/2-1}, \ldots, Y^*_1, Y^*_0 \right]$, where $Y_k = ib_k$ and b_k is the real-valued symbol drawn from a constellation such as PAM. In PAM-DMT, the FT block will perform IFFT operations on input data. The real part of each subcarrier is not modulated. The time-domain real output signal y_m is generated by taking IFFT of the input vector.

$$
\begin{aligned}
y_m &= \frac{1}{N} \sum_{k=0}^{N-1} Y_k e^{j2\pi \frac{k}{N} m} \\
&= \frac{1}{N} \left[\sum_{k=0}^{(N/2)-1} Y_k e^{j2\pi \frac{k}{N} m} + \sum_{k=0}^{(N/2)-1} Y_{N-k} e^{j2\pi \frac{(N-k)}{N} m} \right] \\
&= \frac{1}{N} \left[\sum_{k=0}^{(N/2)-1} i \left(b_k e^{j2\pi \frac{k}{N} m} - b_k e^{-j2\pi \frac{k}{N} m} \right) \right] \\
&= \frac{1}{N} \left[\sum_{k=0}^{(N/2)-1} i \left(b_k e^{j2\pi \frac{k}{N} m} - \left[b_k e^{j2\pi \frac{k}{N} m} \right]^* \right) \right] \\
&= \frac{-2}{N} \left[\sum_{k=0}^{(N/2)-1} b_k \sin\left(2\pi \frac{k}{N} m \right) \right] \quad m = 0,1,2,\ldots N-1
\end{aligned}
\tag{6.11}
$$

y_m is an antisymmetric signal and has the same information in both positive and negative parts. Mathematically,

$$
\begin{aligned}
y_{m=N-s} &= \frac{-2}{N} \left[\sum_{k=0}^{(N/2)-1} b_k \sin\left(2\pi \frac{k}{N}(N-s) \right) \right] \quad s = 0,1,2,\ldots,N-1 \\
&= \frac{2}{N} \left[\sum_{k=0}^{(N/2)-1} b_k \sin\left(2\pi \frac{k}{N} s \right) \right] \\
&= -y_m
\end{aligned}
\tag{6.12}
$$

We can easily clip the negative part of the signal without losing any information. Therefore, after adding a CP to the IFFT output, the negative half of the signal is clipped. Clipping noise is found to be falling over only on the real part of each subcarrier [11]. Thus, because of the orthogonality of clipping noise, transmitted symbols remain uncorrupted by the noise and can be recovered easily at the receiver.

The clipping operation is the same as defined in the previous subsection. The clipped output $\lfloor y_n \rfloor_c$ is passed through a D/A converter to generate continuous time signal which finally modulates the intensity of the optical modulator.

At the receiver side, we perform the reverse operations in a similar fashion to that of ACO-OFDM to extract the useful data. The only difference is that at the output of the TF block which performs FFT operation, we only extract the imaginary part of the first half subcarriers.

The received signal at a specific subcarrier in the absence of any noise is given by

$$
\begin{aligned}
Y_k &= \sum_{m=0}^{(N/2)-1} \lfloor y_m \rfloor_c e^{-j2\pi\frac{k}{N}m} + \sum_{m=0}^{(N/2)-1} \lfloor y_{N-m} \rfloor_c e^{-j2\pi\frac{k}{N}(N-m)} \\
&= \sum_{m=0}^{(N/2)-1} \lfloor y_m \rfloor_c e^{-j2\pi\frac{k}{N}m} + \sum_{m=0}^{(N/2)-1} \lfloor -y_m \rfloor_c e^{-j2\pi\frac{k}{N}(N-m)} \\
&= \sum_{m=0}^{(N/2)-1} \left\{ \lfloor y_m \rfloor_c e^{-j2\pi\frac{k}{N}m} + \lfloor -y_m \rfloor_c e^{j2\pi\frac{k}{N}m} \right\} \\
&= \sum_{m=0}^{(N/2)-1} \left\{ \lfloor y_m \rfloor_c \left[\cos\left(2\pi\frac{k}{N}m\right) - i\sin\left(2\pi\frac{k}{N}m\right) \right] + \lfloor -y_m \rfloor_c \left[\cos\left(2\pi\frac{k}{N}m\right) + \sin\left(2\pi\frac{k}{N}m\right) \right] \right\} \\
&= \sum_{m=0}^{(N/2)-1} \left\{ \left[\lfloor y_m \rfloor_c + \lfloor -y_m \rfloor_c \right] \cos\left(2\pi\frac{k}{N}m\right) - i\left[\lfloor y_m \rfloor_c - \lfloor -y_m \rfloor_c \right] \sin\left(2\pi\frac{k}{N}m\right) \right\} \\
&= \sum_{m=0}^{(N/2)-1} \left\{ |y_m| \cos\left(2\pi\frac{k}{N}m\right) - i[y_m]\sin\left(2\pi\frac{k}{N}m\right) \right\}
\end{aligned}
\tag{6.13}
$$

Equation (6.13) shows that clipping noise falls on the real part of each subcarrier and it actually gives the absolute value of the transmitted time domain signal. This valuable information can be used to improve overall SNR by a few decibel with some additional signal processing.

6.3.3 DHT-OFDM

In DHT-based optical OFDM, a vector of length $N/2$ of real symbols drawn from a real constellation such as M-PAM forms the input to the FT block. In this scheme, the FT block will perform inverse fast Hartley transform (IFHT). According to Ref. [14], if the input symbols only modulate odd indexed subcarriers, clipping noise will only fall on even indexed subcarriers. Therefore, the input vector of length N is transformed to $\mathbf{X} = [0, X_0, 0, X_1, \ldots, X_{N/4-1}, 0, X_{N/4}, \ldots, X_{N/2-1}, 0, X_{N/2}]$ by the zero insertion block. However, we do not need conjugates of the input symbols as IFHT is a real transformation and will generate real signals with real input symbols. Therefore, the length of useful input symbols is $N/2$. An N-point IFHT is performed on $\mathbf{X}$ to output a real bipolar signal:

$$
x(n) = \frac{1}{\sqrt{N}} \sum_{k=0}^{N-1} X(k)\left[\cos(2\pi kn/N) + \sin(2\pi kn/N) \right]
\tag{6.14}
$$

The remaining transmitter front end blocks perform the same operation on this bipolar signal as that in ACO-OFDM and finally transmit it using an optical transmitter.

At the receiver, reverse operation is performed to recover transmitted bits. After removal of CP, fast Hartley transform (FHT) is performed by the TF block on the received signal which outputs estimated transmitted symbols. DHT has a self-inverse property which enables us to use the same software routines as used by the transmitter.

6.4   Precoding and PAPR Reduction in AC OFDM OW Systems

In this section, we have analyzed different precoding-based PAPR reduction techniques for AC optical OFDM wireless communication systems. Due to the nonlinear characteristics of optical transmitters in IM/DD systems such as LED, high PAPR input signals will suffer from distortion due to clipping. OFDM systems suffer from high PAPR problems that can limit its performance in IM/DD systems. Therefore, PAPR reduction techniques have to be employed. Although a large number of PAPR reduction techniques have been proposed for RF-based OFDM systems [15–18], we see only very limited literature about these techniques for optical OFDM communication system. Therefore, in this section we analyze some precoding-based PAPR reduction techniques for ACO-OFDM and PAM-DMT. We have described DFT coding, Zadoff-Chu Transform (ZCT) [19] and discrete cosine transform (DCT) for ACO-OFDM and only DCT for PAM-DMT as the modulating symbols are real. Performance comparison of these precoding techniques is shown using different QAM modulation schemes. Simulation results have shown that both DFT and ZCT offer more PAPR reduction than DCT in ACO-OFDM. For PAM-DMT, DCT precoding yields significant PAPR reduction compared to the conventional PAM-DMT signal. These precoding schemes also offer the advantage of zero signaling overhead.

6.4.1   Precoding-Based Optical OFDM System Model

A block diagram of a baseband precoding-based AC optical OFDM system is shown in Figure 6.4.

In precoding-based OFDM systems, data is transmitted in blocks where each block represents one OFDM symbol. In each block, a parallel stream of N input data symbols $\mathbf{X} = \left[X_0, X_1, \ldots, X_{N-1} \right]^T$, where $X_k = a_k + ib_k$ and a_k is the real part and b_k is the imaginary part, drawn from 2-D constellations such as QPSK, 16- and 64- QAM are first precoded with the precoding scheme giving an output vector $\mathbf{X_p} = \mathbf{PX}$, where $\mathbf{P}$ is $N \times N$ precoding matrix. The precoded output symbols which will modulate the individual subcarriers form the input to the IFFT block. In ACO-OFDM, only odd subcarriers are modulated by the complex input symbols. Even subcarriers are not modulated and are set to zero. A Mapping/ Zero Insertion block performs input vector formatting prior to IFFT to achieve selective subcarrier modulation. Therefore, the input data vector to the IFFT block becomes $\bar{\mathbf{X}} = \left[0, X_{p,1}, 0, X_{p,3}, \ldots, X_{p,N-1}, 0, X_{p,N-1}^*, 0, \ldots, X_{p,1}^* \right]^T$. A real-valued output $x(n)$ is generated by performing 4N-point IFFT on the conjugate symmetric data frame.

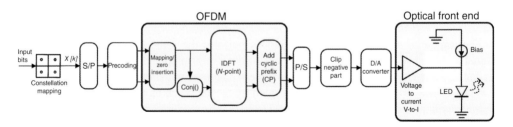

Figure 6.4   Precoding-based optical OFDM system model with clipping.

In PAM-DMT, on the other hand, symbols from a real valued constellation such as M-PAM are used to modulate the complex part of each subcarrier. In precoding-based OFDM systems, the real valued data symbols are first precoded and then they are used to modulate the complex part of each subcarrier. To achieve this, precoded input data vector is formatted by Mapping/ Zero insertion block which gives an output vector $\overline{\mathbf{X}} = \left[0, X_{p,1}, \ldots, X_{p,N-1}, 0, X_{p,N-1}^{*}, \ldots, X_{p,1}^{*} \right]^{\mathrm{T}}$, where $\overline{X}_k = iC_k$ and C_k is the magnitude of the real symbol output from the precoding block. A real-valued output $x(n)$ is generated by performing 2N-point IFFT on the conjugate symmetric input data frame.

The block of parallel real sample outputs from the CP block is converted into a serial discrete-time domain signal by a parallel to serial (P/S) converter. The signal is asymmetrically clipped by clipping the negative part to produce an output signal that is strictly positive. The clipped signal $x_c(n)$ finally modulates the intensity of the optical transmitter. CP is a copy of the last L samples of the OFDM symbol; we will not include this in our simulations as it will not affect the PAPR analysis.

For an LED-based OW transmitter, the clipped input signal has to be DC-biased to operate in the linear region of the LED current–voltage (I–V) curve usually known as transfer characteristics. A typical LED I–V curve is shown in Figures 6.5 and 6.6. The I–V curve shows the relationship between the forward voltage and forward current through LED. The bias point has to be selected carefully in order to keep the signal variations within the linear region. The nonlinear transfer characteristics also show that if the input signal exceeds the linear region, the output current will be clipped, which will distort the signal and generate out-of-band emissions. One solution to this problem is to reduce the overall intensity of the optical transmitter by decreasing the input signal power. This will reduce the SNR at the receiver causing receiver performance degradation. Another solution is to minimize the maximum value of peaks

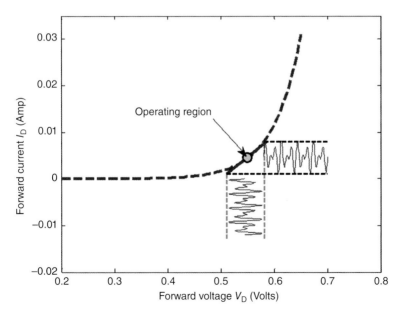

Figure 6.5 A typical LED nonlinear I–V characteristics curve. The curve shows nonlinear relationship between forward current and forward voltage.

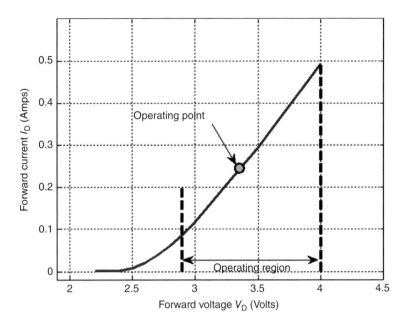

Figure 6.6 Transfer characteristics of OPTEK, OVSPxBCR4 1-W white LED. Typical operating region is between 2.9 and 4 V.

occurring in the OFDM signal envelope without reducing the average power. This will decrease the PAPR of the intensity modulating signal. It can be done by using various PAPR reduction techniques.

PAPR is an important signal parameter that gives an estimate of the envelope variations of the transmitted signal. These envelope variations are critical in the design of RF/Optical transmitter front ends. For OFDM, PAPR is computed over one symbol [0, T] and is defined as follows [15]:

$$PAPR = \frac{\max\limits_{0 \leq n \leq 2N-1} x_c^2(n)}{E\left[x_c^2(n)\right]} \tag{6.15}$$

Another important factor that is related to PAPR and also shows the characteristics of signal envelope is the crest factor (CF) which is also defined over one OFDM symbol [0, T] and is given by

$$CF = \frac{\max\limits_{0 \leq n \leq 2N-1} x_c(n)}{\sqrt{E\left[x_c^2(n)\right]}} \tag{6.16}$$

We calculate the PAPR of the clipped signal $x_c(n)$ and compare it for various complex digital constellations and for various precoding techniques.

6.4.2 *Precoding Schemes*

Although various PAPR reduction methods exist in literature, precoding offers certain advantages over other techniques: it is signal independent, requires comparatively less computational cost, and does not need any signaling overhead. Precoding is a one-shot process wherein the input signal vector is pre-multiplied by a precoding matrix $\mathbf{P}$ given by

$$
\mathbf{P} =
\begin{bmatrix}
a_{0,0} & a_{0,1} & \cdots & a_{0,N-1} \\
a_{1,0} & & & \\
\vdots & \vdots & \vdots & \vdots \\
a_{N-1,0} & & & a_{N-1,N-1}
\end{bmatrix}
\tag{6.17}
$$

where $a_{i,j}$ represents an element of the ith row and jth column of $\mathbf{P}$. This precoding matrix can be generated in different ways, depending on the precoding schemes. We analyze three schemes here.

6.4.2.1 DFT Precoding

In this precoding method, a $N \times N$ precoding matrix is generated that transforms the input data vector to frequency domain. This matrix simply performs an FFT operation and can be generated by

$$
a_{n,k} = \left\{ W_N^{kn}, \quad \text{where } 0 \le k \le N-1, 0 \le n \le N-1 \right.
\tag{6.18}
$$

where n and k are the row and the column index, respectively. W_N^{kn} is the Nth root of unity. This transformation generates a new frequency domain symbol vector of size $N \times 1$ which is obtained by pre-multiplying the input vector by $\mathbf{P}$, that is, $\mathbf{X}_p = \mathbf{PX}$.

6.4.2.2 Zadoff–Chu Sequence Precoding

These sequences are a class of generalized chirp-like sequences that have ideal autocorrelation properties. They also have a property of constant magnitude cross correlation. A Zadoff–Chu sequence of length N is defined by

$$
a_k =
\begin{cases}
e^{j\frac{2\pi r}{N}\left(\frac{k^2}{2}+qk\right)}, & N \text{ even} \\
e^{j\frac{2\pi r}{N}\left(\frac{k(k+1)}{2}+qk\right)}, & N \text{ odd}
\end{cases}
\tag{6.19}
$$

where $k = 0,1,2,\ldots,N-1$ and r is the code index relatively prime to N. q is an integer. A precoding matrix based on Zadoff–Chu sequences of size $N \times N$ can be formed by

$$
\mathbf{P} =
\begin{bmatrix}
a_0 & a_1 & \cdots & a_{N-1} \\
a_N & a_{N+1} & \cdots & a_{2N-1} \\
\vdots & \vdots & \ddots & \vdots \\
a_{(N-1)N} & \cdots & \cdots & a_{N^2-1}
\end{bmatrix}
\tag{6.20}
$$

6.4.2.3   Discrete Cosine Transform (DCT) Precoding

DCT has a very good energy compaction property that makes it very attractive for precoding. Its ability to represent the input signal with very few coefficients will result in an OFDM output signal that has reduced PAPR. This reduction results from the fact that after DCT precoding, the input vector to IFFT block has comparatively fewer high valued elements than the original input. Although several definitions exists for a DCT, we use the most popular one, which is 1-D DCT, given by

$$d_n = \sqrt{\frac{2}{N}} \alpha_n \sum_{k=0}^{N-1} X_k \cos\left(\pi n\left(\frac{2k+1}{2N}\right)\right) \tag{6.21}$$

where

$$\alpha_n = \begin{cases} \dfrac{1}{\sqrt{2}}, & n = 0 \\ 1, & n = 1, 2, \ldots, N-1 \end{cases}.$$

A $N \times N$ DCT precoding matrix $\mathbf{P}$ can be obtained from

$$P_{n,k} = \sqrt{\frac{2}{N}} \alpha_n \cos\left(\pi n\left(\frac{2k+1}{2N}\right)\right) \tag{6.22}$$

where $0 \le n \le N-1$ is the row and $0 \le k \le N-1$ is the column index. The output of the precoder is a $N \times 1$ coded vector $\mathbf{X}_p$.

In case of ACO-OFDM, the input can be a complex data vector with real component denoted by $\mathbf{X}_{real}$ and imaginary component represented by $\mathbf{X}_{imag}$. In this case, the precoded output is given by $\mathbf{X}_p = \text{DCT}\left(\mathbf{X}_{real}\right) + \text{DCT}\left(\mathbf{X}_{imag}\right)$.

For PAM-DMT, as the input data symbol vector $\mathbf{X}$ contains only real components as they are drawn from a real mapping scheme such as M-PAM, DCT precoding is one of the suitable schemes that output real frequency coefficients. The precoding operation in matrix form can be written as $\mathbf{X}_p = \mathbf{PX}$. The components of the output vector are purely real, which is in contrast to the other precoding schemes. These real precoded data symbols finally modulate the imaginary parts of each subcarrier in OFDM. This is accomplished by the mapping/zero-insertion block.

6.4.3   Simulation Results and Discussions

In this section, we compare the performance of various precoding techniques in reducing the PAPR for two types of clipping-based OW OFDM systems. In case of ACO-OFDM, input data vector of length $N = 128$ is generated by drawing symbols from QPSK, 16- and 64-QAM and $4N = 512$ point IFFT is used to generate the output OFDM signal. For PAM-DMT, input data symbol vector of length $N = 128$ is formed by drawing symbols from M-PAM, where $M = 4$, 16 and 64 and IFFT size of $2N = 256$ is used to generate the output time domain signal [20].

PAPR performance is usually shown using complementary cumulative distribution function (CCDF) curves. These curves show the probability that PAPR is higher than a specified $PAPR_0$, that is, $Pr(PAPR > PAPR_0)$. These curves are obtained through extensive MATLAB simulations by generating random input data for the different constellations.

Figures 6.7 and 6.8 show the CCDF comparison of PAPR for ACO-OFDM and DFT precoded ACO-OFDM. The curves show that DFT precoding reduces the PAPR of ACO-OFDM signal by a few decibels.

Figures 6.9 and 6.10 show the CCDF comparison of PAPR for ACO-OFDM and DCT precoded ACO-OFDM. The curves show that with DCT precoding, we see significant reduction in the PAPR of ACO-OFDM signals.

Figures 6.11 and 6.12 show the CCDF comparison of PAPR for ACO-OFDM and ZC sequence precoded ACO-OFDM. The curves show that ZC sequences reduce the PAPR of the AC OFDM signal by approximately 3 dB at clipping level of 10^{-4} and thus prove to be a promising PAPR reduction precoding technique.

Figures 6.13 and 6.14 show the CCDF curves for PAM-DMT and DCT precoded PAM-DMT for different digital constellations. The curve shows that with DCT precoding, the PAPR of AC OFDM signal is reduced by approximately 3 dB at clipping level of 10^{-4}. Therefore, the DCT precoding scheme definitely proves to be a strong candidate for PAPR reduction for PAM-DMT.

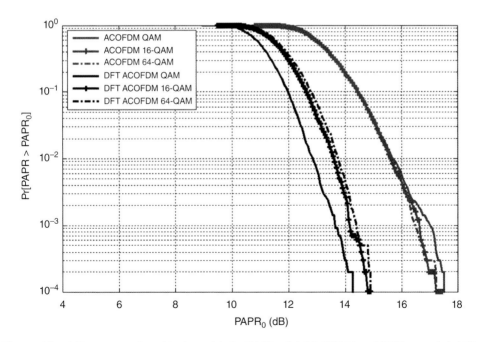

Figure 6.7 CCDF curves (logarithmic scale) for PAPR of ACO-OFDM and DFT precoded ACO-OFDM for 4-, 16-, and 64-QAM (logarithmic scale).

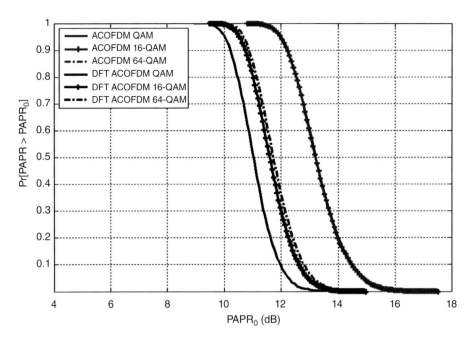

Figure 6.8 CCDF curves for PAPR of ACO-OFDM and DFT precoded ACO-OFDM for 4-, 16-, and 64-QAM.

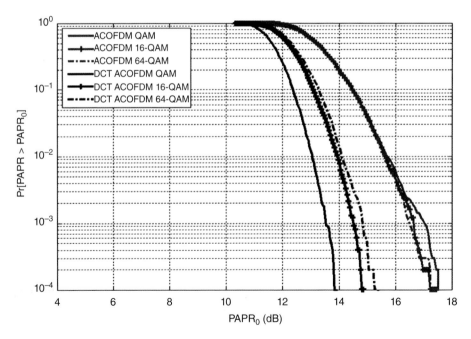

Figure 6.9 CCDF curves (logarithmic scale) for PAPR of ACO-OFDM and DCT precoded ACO-OFDM for 4-, 16-, and 64-QAM.

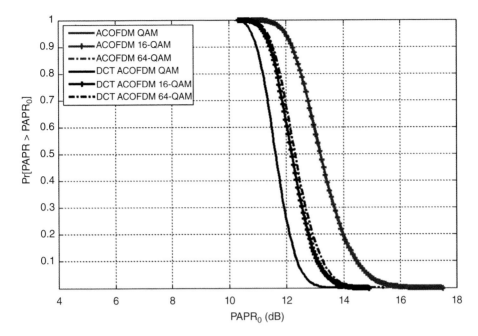

Figure 6.10　CCDF curves for PAPR of ACO-OFDM and DCT precoded ACO-OFDM for 4-, 16-, and 64-QAM.

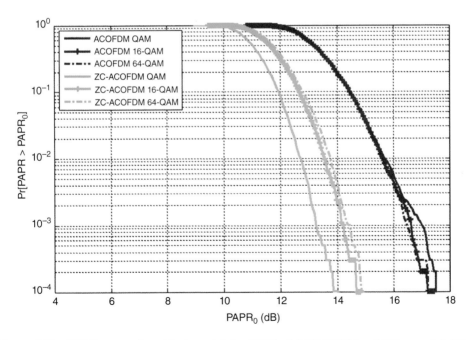

Figure 6.11　CCDF curves (logarithmic scale) for PAPR of ACO-OFDM and ZC precoded ACO-OFDM for 4-, 16-, and 64-QAM.

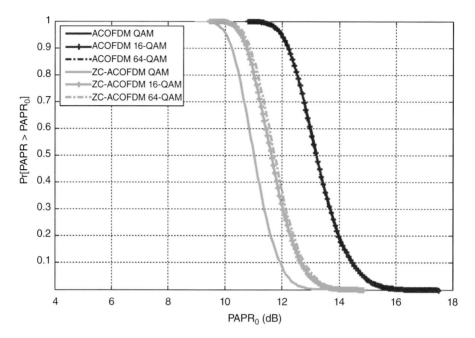

Figure 6.12 CCDF curves for PAPR of ACO-OFDM and ZC precoded ACO-OFDM for 4-, 16-, and 64-QAM.

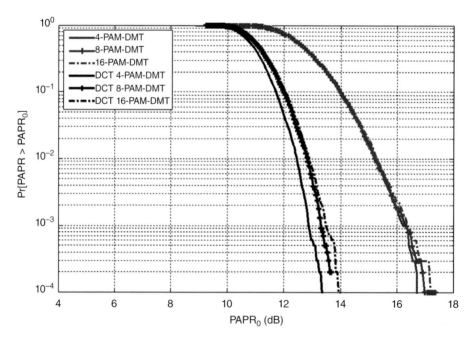

Figure 6.13 CCDF curves (logarithmic scale) for PAPR of PAM-DMT and DCT precoded PAM-DMT for 4-, 8-, and 16-PAM.

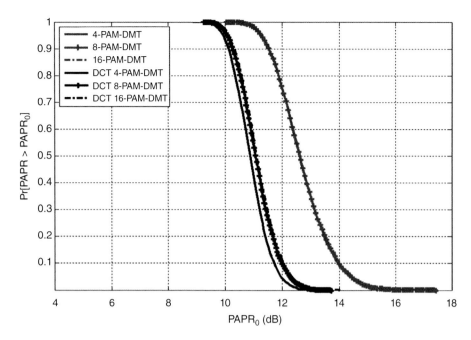

Figure 6.14 CCDF curves for PAPR of PAM-DMT and DCT precoded PAM-DMT for 4-, 8-, and 16-PAM.

6.5 Performance of AC OFDM Systems in AWGN and Multipath Channel

In this section, we compare BER performance of precoding-based ACO-OFDM and PAM-DMT OW systems in AWGN and indoor multipath channels. Simulation and analytical results show that precoding schemes such as DFT, DC and ZC sequence do not affect the performance of the OW systems in AWGN channel, while they reduce the PAPR of OFDM output signal. However, in multipath indoor channel, by using zero-forcing frequency-domain equalization (ZF-FDE), precoding-based systems give better BER performance than their conventional counterparts. With additional clipping to further reduce the PAPR, precoding-based systems also show better BER performance compared to nonprecoded systems when clipped, relative to the peak of nonprecoded systems. Therefore, precoding-based ACO-OFDM and PAM-DMT systems offer better BER performance with zero signaling overhead and low PAPR, compared to the conventional systems.

6.5.1 Precoding-Based OW OFDM System Model with AWGN

A block diagram of AC optical OFDM system with precoding is shown in Figure 6.15. We use a discrete time baseband system model for both ACO-OFDM and PAM-DMT. In ACO-OFDM, a vector of M input symbols drawn from a complex constellation such as M-ary QAM forms input to precoding block. A precoding matrix $\mathbf{P}$ transforms these input symbols to precoded output $\mathbf{Y} = \mathbf{PX}$. These precoded symbols only modulate the odd subcarriers. A discrete

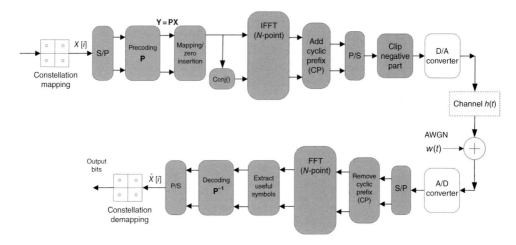

Figure 6.15   A baseband AC-based optical OFDM system diagram.

time output signal is generated by the IFFT block. This time domain signal is then asymmetrically clipped to generate a unipolar signal. This unipolar signal propagates through the channel $h(n)$ and is detected by an optical detector such as a photodiode, which converts it to an electrical signal $z(t)$. The received signal is given by

$$z(t) = x_c(t) * h(t) + w(t) \tag{6.23}$$

where $w(t)$ represents discrete time samples of AWGN and * represents convolution operation. The received signal is then sampled by an A/D converter to obtain a discrete-time signal $x_r(n)$. The corresponding discrete version of the channel impulse response can be represented by $h(n)$. In case of AWGN channel,

$$h(n) = \begin{cases} 1 & n = 0 \\ 0 & n \neq 0 \end{cases} \tag{6.24}$$

The precoded symbols are decoded by using the inverse of the precoding matrix to obtain estimated symbols. The estimated symbols are compared with transmitted symbols to get BER performance of the system.

In PAM-DMT on the other hand, a vector of N input symbols drawn from a real constellation such as M-ary PAM modulates the complex part of each subcarrier. As we need real input symbols for each subcarrier, for PAM-DMT we use only DCT precoding, which will yield real output coefficients for a real input vector.

6.5.2   Multipath Indoor Channel

Several techniques have been proposed to numerically generate the impulse response of an indoor multipath channel [21, 22] for OW systems. We follow the method used in Ref. [21] in our simulations. In our implementation, we placed the source at the ceiling of a room with dimensions $5\,m \times 5\,m \times 3\,m$ pointing downwards and the receiver facing upwards at a height of

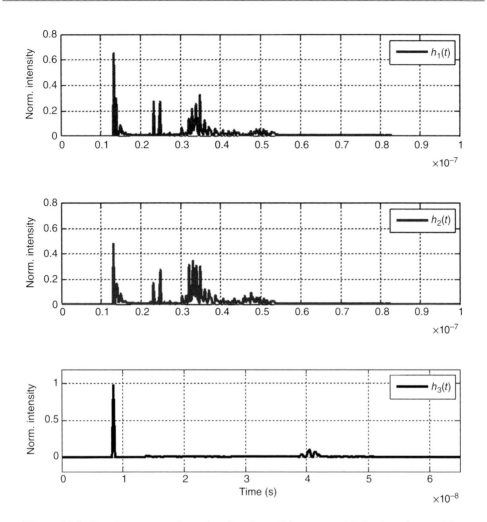

Figure 6.16 Impulse response for various locations of the source with fixed receiver position.

1 m from the floor. The room surfaces are assumed to be diffusive in nature. The receiver FOV is assumed to be 60° and detector area to be $1 \times 10^{-4}\,\text{m}^2$. We only consider light beams with at most three reflections and the beams arriving directly through LOS. Figure 6.16 shows sample impulse responses generated by changing the source location at three different places on the ceiling. The first peak shows the LOS component.

6.5.3   *Frequency-Domain Equalization (FDE)*

One of the advantages of OFDM-based systems is the use of FDE [23, 24]. In this study, we assume that we have perfect knowledge of the channel, and we use this information to equalize received signal in the frequency domain.

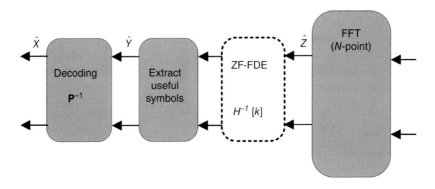

Figure 6.17   ZF-FDE for precoding-based ACO-OFDM and PAM-DMT.

We denote $H(k)$ as the channel transfer function value at subcarrier index k. In a simple linear equalization scheme in which we try to nullify the channel effect on the information symbol $X(k)$, the equalization coefficient at subcarrier index k turns out to be $C(k) = H(k)^{-1}$. This is called zero-forcing (ZF) equalization. For OFDM-based OW system, ZF-FDE at the receiver is illustrated in Figure 6.17. Because of its simplicity, we use ZF-FDE in our simulations to evaluate the performance of ACO-OFDM and PAM-DMT systems.

In case of ACO-OFDM, symbols from FFT output are simply multiplied by the equalizer coefficients and useful symbols are extracted. However, in PAM-DMT, as symbols are drawn from real constellation, the FFT output complex symbols should first be multiplied with the equalizer coefficient and then the complex part of the equalized symbols extracted to obtain estimated PAM symbols.

6.5.4   Analytical BER Performance Results

In this section, we derive analytical results for BER performance of precoding-based OW system. We assume that AWGN has variance $\sigma_n^2 = N_0$ and the total average power of the channel impulse response is unity, that is, $\sum_{K=0}^{M-1}|H(K)|^2 = 1$. We also consider total average transmitted electrical power before clipping to be $E\left[|x(n)|^2\right] = 1$. From (6.23) and Figure 6.17, received symbol at the output of FFT at a specific subcarrier index K is given by

$$Z[k] = X_c[k]H[k] + W[k] \qquad (6.25)$$

The noise variance at the output of the FFT block will not change because of linearity of FFT operation. Due to asymmetric clipping at the transmitter, the power of each transmitted symbol becomes half. In order to scale the power of each symbol to its original value, we simply scale $Z[k]$ by a factor of 2. This will increase noise variance to $\hat{\sigma}_n^2 = 4N_0$. After ZF-FDE, we get

$$\hat{Y} = X[k] + W[k]H[k]^{-1} \qquad (6.26)$$

The noise variance at the output of ZF-FDE becomes $\sigma_{n,\text{FDE}}^2 = \hat{\sigma}_n^2 / |H[k]|^2$. For ACO-OFDM and PAM-DMT systems without precoding, the electrical symbol energy-to-noise power ratio at the output of ZF-FDE is given by

$$\frac{E_{s,\text{elec}}}{N_0} = \frac{1}{\hat{\sigma}_n^2 / |H[k]|^2} \tag{6.27}$$

We see from the above equation that electrical symbol energy-to-noise power ratio depends on subcarrier channel power. In case of M-QAM constellation mapping for ACO-OFDM system, average symbol error rate (SER) for a specific subcarrier is given by [25],

$$P_s \approx \frac{1}{M} 4\left(1 - \frac{1}{\sqrt{M}}\right) Q\left(\sqrt{\frac{E_{s,\text{elec}}}{N_0}}\right) \tag{6.28}$$

However, in case of PAM-DMT, we extract only the complex part of each subcarrier. In case of M-PAM modulation for PAM-DMT system, the average SER is given by

$$P_s = \frac{1}{M} \frac{2(M-1)}{M} Q\left(\sqrt{\frac{E_{s,\text{elec}}}{N_0}}\right) \tag{6.29}$$

By using gray coding, BER becomes $P_b = P_s / \log_2 M$. In a precoded system, the equalized symbols are multiplied by the inverse of the precoding matrix used at the transmitter. Therefore the noise variance at the output of the decoding matrix is given by $\sigma_{n,\text{decode}}^2 = \frac{1}{M} \sum_{k=0}^{M-1} \hat{\sigma}_n^2 / |H[k]|^2$. This is due to the fact that the noise samples remain uncorrelated due to the unitary property of precoding matrices such as DFT and DCT [26]. This expression shows that the noise variance at the output of the decoding matrix for each index will be the same. Therefore, the SER for ACO-OFDM system using M-QAM constellation becomes

$$P_{s,\text{ACO}} \approx 4\left(1 - \frac{1}{\sqrt{M}}\right) Q\left(\sqrt{\frac{E_{s,\text{elec}}}{\frac{1}{M} \sum_{k=0}^{M-1} \hat{\sigma}_n^2 / |H[k]|^2}}\right) \tag{6.30}$$

Similarly, for PAM-DMT, the SER for M-PAM constellation is given by

$$P_{s,\text{PAM}} = \frac{2(M-1)}{M} Q\left(\sqrt{\frac{E_{s,\text{elec}}}{\frac{1}{M} \sum_{k=0}^{M-1} \hat{\sigma}_n^2 / |H[k]|^2}}\right) \tag{6.31}$$

From the above two equations, we observe that in case of AWGN channel, $|H[K]| = 1, K = 0,1,2,\ldots,M-1$, therefore BER performance for ACO-OFDM becomes $P_{s,\text{ACO}} \approx 4\left(1 - (1/\sqrt{M})\right) Q\left(\sqrt{E_{s,\text{elec}}/\hat{\sigma}_n^2}\right)$ and for PAM-DMT $P_{s,\text{PAM}} = (2(M-1)/M) Q\left(\sqrt{E_{s,\text{elec}}/\hat{\sigma}_n^2}\right)$. This shows that in case of precoding in AWGN channel, system performance does not change and is the same as that without precoding.

6.5.5 Electrical and Optical Performance Metrics

In this section, we investigate the impact of electrical to optical conversion of output OFDM signal on the BER performance. To make a fair comparison between precoded and nonprecoded systems, we use normalized $E_{b(opt)}/N_0$ where the average output optical power is set to unity, that is, $E[x_c(n)] = 1$. We obtain values of required normalized $E_{b(opt)}/N_0$ for which BER is 10^{-4} represented by $\langle E_{b(opt)}/N_0 \rangle_{BER}$. We plot our results for various values of bit rate/normalized bandwidth. The bandwidth is normalized with respect to on–off Keying (OOK) and is defined as the location of first spectral null. Therefore, for ACO-OFDM, the bit rate/normalized bandwidth is given by $\log_2(M)/2/(1+2/N)$. In ACO-OFDM, only 1/4 of the total subcarriers carry data, excluding DC and $N/2$-nd subcarrier, therefore the factor 1/2 appears in the above expression. M represents the M-ary QAM constellation size. For PAM-DMT, the spectral null appears at the same location $(1+2/N)$ as that of ACO-OFDM. However, as 1/2 of its total subcarriers carry data excluding DC and $N/2$-nd subcarrier, the bit rate/normalized bandwidth is given by $\log_2(M)/(1+2/N)$ where M is the constellation size of M-ary PAM.

6.5.6 Clipping and PAPR Reduction

PAPR gives a measure of the signal variations relative to the average power. To efficiently transmit the signal using an LED with nonlinear I–V characteristics, we need to have a lower PAPR. In order to further improve PAPR of AC signals, a simple clipping technique can be used. With clipping, we can bias LED at higher values resulting in higher intensity signal and higher average output power. This will increase received SNR. However, due to clipping, BER performance will deteriorate and degradation will depend upon the amount of clipping. To see BER performance variation due to clipping for precoded and nonprecoded systems, we clip the signals relative to the peak of nonprecoded ACO-OFDM and PAM-DMT signal, respectively, for specific signal constellations.

$$V_{clip}(dB) = 20\log_{10}\left(\frac{V_{clip}}{V_{peak}}\right) \tag{6.32}$$

For a given dB amount of clipping, the clipping level of the output signal is chosen as

$$V_{clip} = V_{peak} \times 10^{\frac{V_{clip}(dB)}{20}} \tag{6.33}$$

where V_{clip} is the voltage level at which the output signal is clipped and V_{peak} is the peak value of the non-precoded output signal for the same constellation. The clipping operation can be defined as

$$x_{clip}(n) = \begin{cases} x_c(n) & \text{if} \quad x_c(n) \leq V_{clip} \\ V_{clip} & \text{if} \quad x_c(n) > V_{clip} \end{cases} \tag{6.34}$$

The clipped output signal modulates the intensity of the optical transmitter. This criterion of clipping is useful in choosing a specific bias point of an LED transmitter. Due to precoding, PAPR of the output signal is already reduced and we see fewer peaks. Therefore the effect of clipping on the BER performance of precoded signal is less than on that of conventional AC signal.

Table 6.1 List of parameters to generate multipath impulse response

Room dimensions	$5\,m \times 5\,m \times 3\,m$
Reflectivity of each surface	Ceiling: 0.9; walls: 0.8; floor: 0.3
Receiver location	$(2.5\,m, 2.5\,m, 1\,m)$
Detector area	$1 \times 10^{-4}\,m^2$
Detector FOV	$60°$
Source location (H)	$(0.1\,m, 0.1\,m, 3\,m)$
Source location (M)	$(0.1\,m, 0.2\,m, 3\,m)$
Source location (L)	$(1\,m, 2\,m, 3\,m)$
Source half-power angle	$60°$

6.5.7 Simulation Results

In this section, we present BER performance results for various precoding schemes used in ACO-OFDM and PAM-DMT systems. Extensive MATLAB-based Monte Carlo simulations were performed to obtain the results [27]. For ACO-OFDM, an input symbol vector of length $N = 128$ was generated by drawing symbols from 4-, 16-, and 64-QAM and $4N$-point IFFT was performed to get time domain output OFDM signal. In case of PAM-DMT, a vector of $N = 256$ real symbols drawn from M-PAM was formed where $M = 4$, 8 and 16 and a $2N$-point IFFT was used to generate the output time domain signal. OFDM output sampling rate of $R_s = 400\,Msamp\,s^{-1}$ was chosen for both ACO-OFDM and PAM-DMT. The symbol rate for ACO-OFDM was $R_{ACO} = 94\,MHz$ and $R_{PAM-DMT} = 178\,MHz$ for PAM-DMT. The CP length of $N_{CP} = 32$ was used and was chosen to be always greater than the maximum delay spread of the worst possible channel. To compute the PAPR of the OFDM output signal, an oversampling rate of 4 was used for precise calculation. In order to simulate the multipath channel, we used a room with a source on the ceiling and receiver at 1 m height from the floor. Details of simulation parameters are listed in Table 6.1.

6.5.7.1 Performance of Precoding Schemes in AWGN

Figure 6.18 shows BER performance curves for ACO-OFDM, DCT, DFT, and ZC sequence precoded ACO-OFDM for 4-, 16-, 64-, 256-, and 1024-QAM in an AWGN channel. From the figure, we see that BER performance of conventional ACO-OFDM and precoded systems for respective QAM constellations almost overlap each other. Therefore, precoding does not affect the BER performance in AWGN channel, which proves our analytical result.

Figure 6.19 shows BER performance of PAM-DMT and DCT precoded PAM-DMT in AWGN channel. We observe a similar trend that precoding does not affect system performance.

6.5.7.2 Performance of Precoding Schemes in Multipath Indoor Channel

In this section, we present BER performance results for ACO-OFDM and PAM-DMT systems in multipath indoor channel with ZF-FDE. We plot the variation of $\langle E_{b(opt)}/N_0 \rangle_{BER}$ for various values of bit rate/normalized bandwidths. Figure 6.20a shows BER performance of ACO-OFDM in multipath channel $h_1(t)$ with severe multipath and long delay spread. We observe

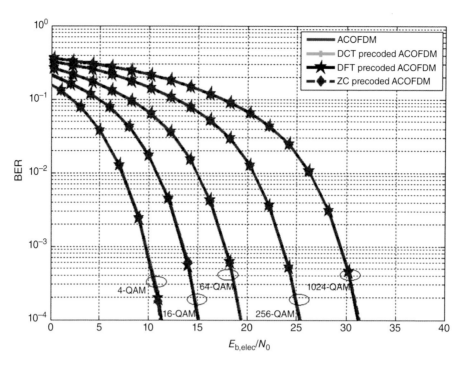

Figure 6.18 BER performance of ACO-OFDM, DCT-, DFT-, and ZC-precoded ACO-OFDM in AWGN channel.

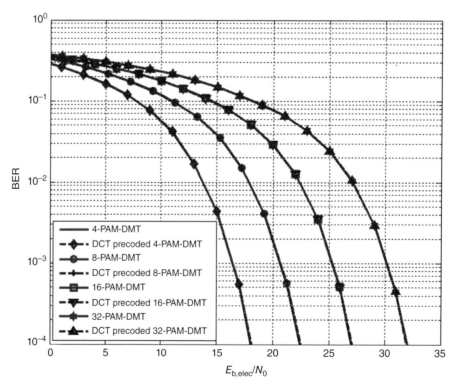

Figure 6.19 BER performance of PAM-DMT and DCT precoded PAM-DMT for 4-, 8-, 16-, and 32-PAM in AWGN channel.

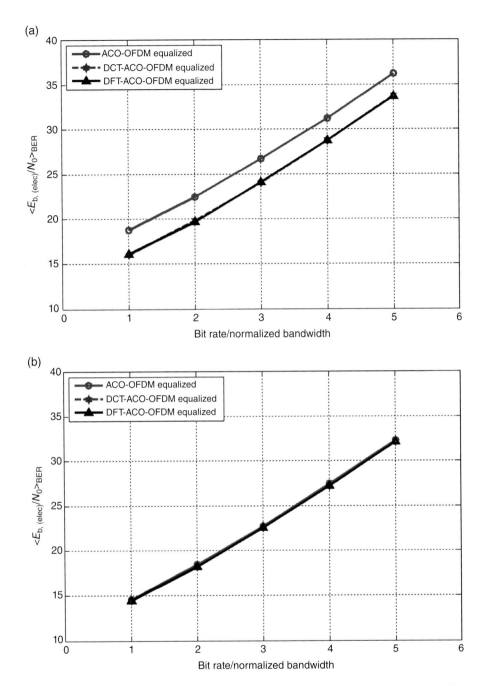

Figure 6.20 Electrical bit energy-to-noise power ratio required for BER of 10^{-4} for ACO-OFDM in multipath channel with ZF-FDE equalization for (a) $h_1(t)$ and (b) $h_3(t)$.

that by using ZF-FDE, precoding improves performance by 3 dB compared to nonprecoded system. We also observe that both DCT and DFT precoding result in the same performance. The performance improvement is due to the fact that with precoding, the SNR for each subcarrier at the output of the decoder is the same because of the averaging effect of decoder matrix. However, in a system without precoding, SNR varies for each subcarrier.

Figure 6.20b shows the BER performance of ACO-OFDM with FDE in multipath channel $h_3(t)$ which has few multipaths and a strong LOS component. We see that the performance of precoded and nonprecoded systems is almost same.

Figure 6.21a and b shows the BER performance of ACO-OFDM in multipath indoor channel $h_1(t)$ and $h_3(t)$, respectively, with ZF-FDE when average optical power was set to unity. The results again show that even in optical domain, the precoding-based systems perform better than or similar to their nonprecoded counterparts.

Figure 6.22a shows electrical bit energy-to-noise power ratio required for BER of 10^{-4} for PAM-DMT in multipath channel $h_1(t)$ with severe multipath and long delay spread. We observe that in the presence of ZF-FDE, precoding gives better BER performance than nonprecoded system. We see a consistent 3 dB performance improvement with precoding.

Figure 6.22b shows electrical bit energy-to-noise power ratio required for BER of 10^{-4} for PAM-DMT in multipath channel $h_3(t)$ with fewer multipath and strong LOS component. We observe that in this case, both systems show similar performance.

Figure 6.23a shows optical bit energy-to-noise power ratio required for BER of 10^{-4} for PAM-DMT in multipath channel $h_1(t)$ with severe multipath and long delay spread. We observe that in the presence of ZF-FDE, precoding gives better BER performance than nonprecoded system. Again we see a performance difference of 3 dB between precoded and nonprecoded systems.

Figure 6.23b shows optical bit energy-to-noise power ratio required for BER of 10^{-4} for PAM-DMT in multipath channel $h_3(t)$. In this case, both systems show the same performance. Therefore, in case of low multipath, both precoded and conventional ACO-OFDM and PAM-DMT systems show identical performance.

6.5.7.3   Performance of Precoding Schemes with Clipping

Figure 6.24a shows BER performance of ACO-OFDM with additional clipping at the front end. Results show that by clipping the unipolar signal 3 dB relative to the peak, we can achieve a sufficient reduction in PAPR as shown in Figure 6.24b without significantly suffering from BER performance degradation. However, to further reduce the PAPR, we can clip the output signal by 6 dB with noticeable degradation in BER.

Similarly, we see that by precoding the input symbols with DCT and clipping the output unipolar signal by 3 dB relative to the nonprecoded unipolar ACO-OFDM signal peak, we see no BER performance degradation and PAPR difference compared to simple DCT precoded system. This is due to the fact that the precoded output signal has less spikes and the average signal peak level is less than that of conventional ACO-OFDM.

However, if we clip signal by 6 dB, we see BER performance degradation but with decrease in PAPR of the output signal. Figure 6.25a and b show results for clipping-based DCT precoded ACO-OFDM system in AWGN channel. We see a similar trend in the BER performance and PAPR reduction in clipping-based DFT and ZC sequence precoded ACO-OFDM systems.

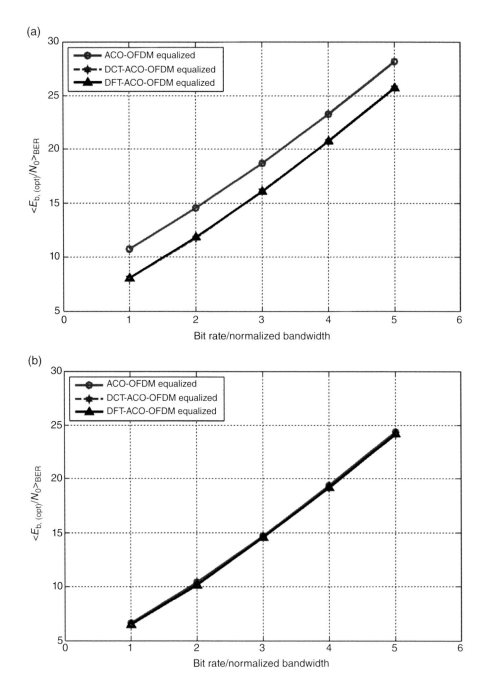

Figure 6.21 Optical bit energy-to-noise power ratio required for BER of 10^{-4} for ACO-OFDM in multipath channel with ZF-FDE equalization for (a) $h_1(t)$ and (b) $h_3(t)$.

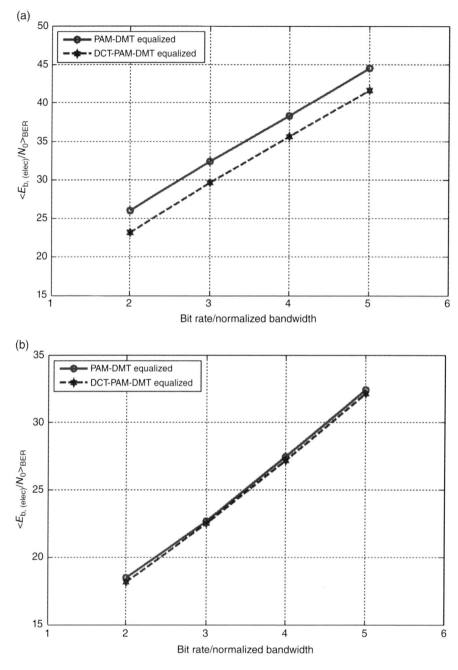

Figure 6.22 Electrical bit energy to noise power ratio required for BER of 10^{-4} for PAM-DMT in multipath channel with ZF-FDE equalization for (a) $h_1(t)$ and (b) $h_3(t)$.

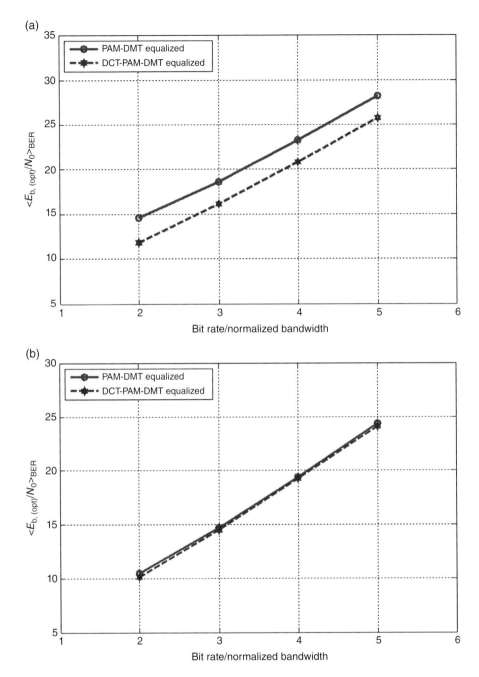

Figure 6.23 Optical bit energy-to-noise power ratio required for BER of 10^{-4} for PAM-DMT in multi-path channel with ZF-FDE equalization for (a) $h_1(t)$ and (b) $h_3(t)$.

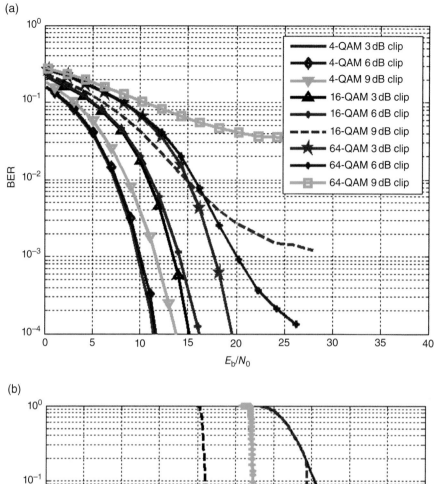

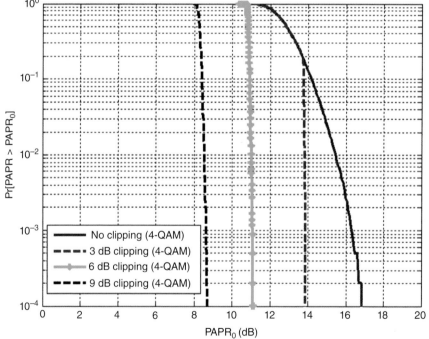

Figure 6.24 BER and PAPR performance of ACO-OFDM with additional clipping in AWGN channel. (a) BER performance and (b) PAPR for 4-QAM.

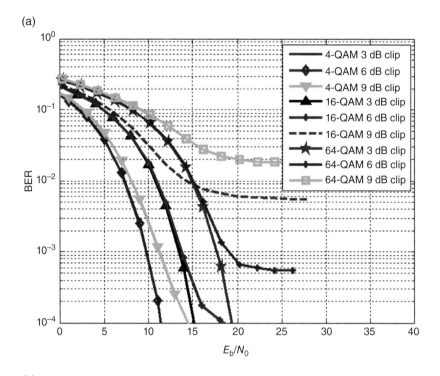

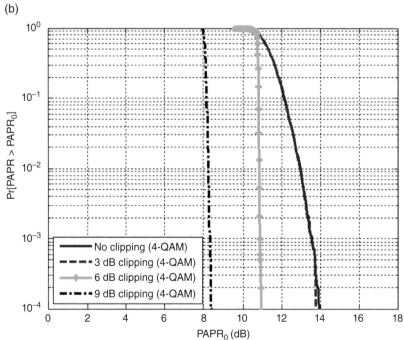

Figure 6.25 BER and PAPR performance of DCT precoded ACO-OFDM with additional clipping in AWGN channel. (a) BER performance and (b) PAPR for 4-QAM.

Figure 6.26a shows BER performance of unipolar PAM-DMT with clipping at the front end. Results show that by clipping the signal 3 dB relative to the peak, we can achieve a sufficient reduction in PAPR as shown in Figure 6.26b without significantly suffering from BER performance degradation. However, to further reduce the PAPR, we can clip the output signal by 6 dB with noticeable degradation in BER.

Figure 6.27a shows the BER performance of DCT precoded PAM-DMT for different clipping levels relative to the peak of conventional PAM-DMT. We see that the BER performance is not severely affected when using 3 dB clipping compared to that of simple PAM-DMT. This shows that we can bias the optical transmitter at least 3 dB higher when using DCT precoding and still achieve the same BER performance. This will enable us to transmit higher average power and get better SNR at the receiver. However, by clipping more than 3 dB, we see significant degradation in BER performance. Figure 6.27b shows the PAPR curves for DCT precoded PAM-DMT scheme. We see that by simply clipping the signal by a few decibels, we can achieve sufficient PAPR reduction without severely degrading the BER performance.

6.6 Conclusions

In this chapter, we have analyzed various precoding techniques for PAPR reduction in clipped OW OFDM systems. We have used DFT precoding, Zadoff–Chu sequence precoding and DCT precoding techniques for ACO-OFDM and PAM-DMT systems. Both of these systems use asymmetric clipping to make the intensity modulating signal positive. We have observed that for ACO-OFDM, Zadoff–Chu precoding gives the maximum PAPR reduction of about 3 dB. In case of PAM-DMT, DCT precoding reduces the PAPR of intensity modulating signal by about 3 dB compared to uncoded PAM-DMT. These precoding schemes besides reducing the PAPR also offer advantages such as signal independence, low computational complexity, and zero signaling overhead. All these advantages and benefits make precoding one of the most desirable PAPR reduction techniques.

We have also compared the BER and PAPR performance of ACO-OFDM, precoding-based ACO-OFDM, PAM-DMT and precoding-based PAM-DMT in AWGN and multipath indoor channel environments. Simulation results show that in AWGN channel, the BER performance curves for precoding-based ACO-OFDM and PAM-DMT are almost identical to the conventional ACO-OFDM and PAM-DMT respectively.

We also observed that precoding reduces PAPR of the output unipolar signal and PAPR can be further reduced with additional clipping at the front end at the cost of some degradation in BER performance, which depends on the amount of clipping. However, the effect of clipping on BER was not severe for precoding-based schemes as it showed better BER performance than their conventional counterparts. Therefore, precoding-based ACO-OFDM and PAM-DMT offer better BER and PAPR performance as compared to the conventional schemes when clipped at the same level, relative to the peak of the nonprecoded schemes.

In case of multipath channel, we observed that the performance of both systems severely degrades in case of higher delay spread. However, when we have perfect knowledge of channel impulse response, by using ZF-FDE, the precoding-based systems perform 3 dB better than their conventional counterparts. Simulation results show the same trend both in electrical and

(a)

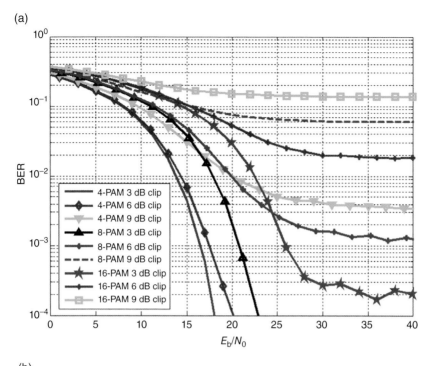

(b)

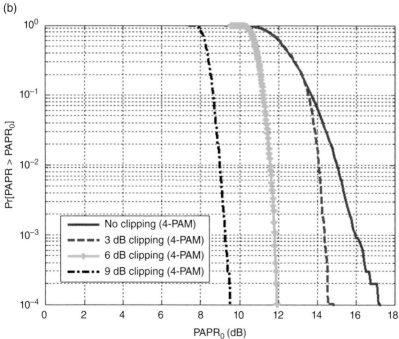

Figure 6.26   BER and PAPR performance of PAM-DMT with additional clipping in AWGN channel.
(a) BER performance and (b) PAPR for 4-PAM.

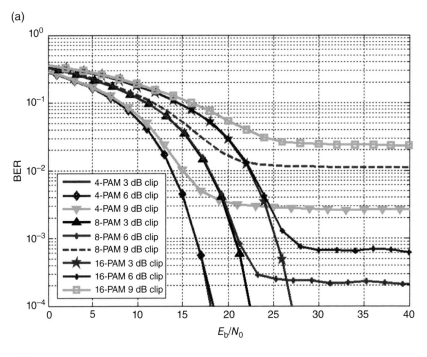

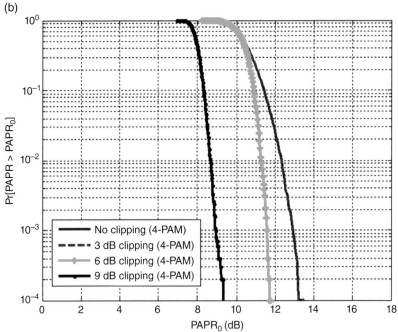

Figure 6.27 BER and PAPR performance of DCT precoded PAM-DMT with additional clipping in AWGN channel. (a) BER performance and (b) PAPR for 4-PAM.

optical domains. Therefore, it can be seen that precoding not only improves PAPR but also offers better performance in multipath indoor channel environments. Hence, precoding can be a promising technique for future OFDM OW systems.

References

[1] L. J. Cimini, "Analysis and simulation of a digital mobile channel using orthogonal frequency division multiplexing," *IEEE Transactions on Communications*, vol. 33, pp. 665–667, 1985.

[2] J. Zyren, "Overview of 3GPP Long Term Evolution LTE physical layer," White Paper, Freescale, 2007.

[3] M. Z. Afgani, H. Haas, H. Elgala, and D. Knipp, "Visible light communication using OFDM," in 2nd International Conference on Testbeds and Research Infrastructure for the Development of Networks and Communities, 2006, Barcelona, Spain, March, 2006.

[4] H. Elgala, R. Mesleh, and H. Haas, "Practical considerations for indoor wireless optical system implementation using OFDM," in Proceedings of ConTEL, pp. 25–30, 2009, Zagreb, Croatia, June, 2009.

[5] J. Armstrong, "OFDM for optical communications," *Journal of Lightwave Technology*, vol. 27, pp. 189–204, 2009.

[6] M. Kavehrad, "Broadband room service by light," *Scientific American*, vol. 297, pp. 82–87, 2007.

[7] J. M. Kahn, J. M. Barry, M. D. Audeh, J. B. Carruthers, W. J. Krause, and G. W. Marsh, "Non-directed infrared links for high-capacity wireless LANs," *IEEE Personal Communications*, vol. 1, no. 2, pp. 12–25, 1994.

[8] J. B. Carruthers and J. M. Kahn, "Multiple-subcarrier modulation for non-directed wireless infrared communication," *IEEE Journal on Selected Areas in Communications*, vol. SAC-14, pp. 538–546, 1996.

[9] H. Elgala, R. Mesleh, and H. Haas, "A study of LED nonlinearity effects on optical wireless transmission using OFDM," in IFIP International Conference on Wireless and Optical Communications Networks, WOCN'09, 2009, Cairo, Egypt, April, 2009.

[10] O. Gonzalez, R. Perez-Jimenez, S. Rodriguez, J. Rabadan, and A. Ayala, "Adaptive OFDM system for communications over the indoor wireless optical channel," *IEE Proceedings Optoelectronics*, vol. 153, no. 4, pp. 139–144, 2006.

[11] S. C. Jeffrey, S. Randel, F. Breyer, and A. M. J. Koonen, "PAM-DMT for intensity modulated and direct-detection optical communication systems," *IEEE Photonics Technology Letters*, vol. 21, no. 23, pp. 1749–1751, 2009.

[12] J. Armstrong and B. J. C. Schmidt, "Comparison of asymmetrically clipped optical OFDM and DC-biased Optical OFDM in AWGN," *IEEE Communications Letters*, vol. 12, no. 5, pp. 343–345, 2008.

[13] J. Armstrong, B. J. C. Schmidt, D. Kalra, H. A. Suraweera, and A. J. Lowery, "Performance of asymmetrically clipped optical OFDM in AWGN for an intensity modulated direct detection system," in Proceedings of GLOBECOM, pp. 1–5, 2006, San Francisco, CA.

[14] M. S. Moreolo, R. Munoz, and G. Junyent, "Novel power efficient optical OFDM based on hartley transform for intensity-modulated direct-detection systems," *Journal of Lightwave Technology*, vol. 28, no. 5, pp. 798–805, 2010.

[15] S. H. Han and J. H. Lee, "An overview of peak-to-average power ratio reduction techniques for multicarrier transmission," *IEEE Wireless Communication*, vol. 12, no. 2, pp. 56–65, 2005.

[16] S. H. Müller and J. B. Huber, "OFDM with reduced peak–to–average power ratio by optimum combination of partial transmit sequences," *Electronics Letters*, vol. 33, no. 5, pp. 368–369, 1997.

[17] R. W. Bäuml, R. F. H. Fisher, and J. B. Huber, "Reducing the peak-to-average power ratio of multicarrier modulation by selected mapping," *Electronics Letters*, vol. 32, no. 22, pp. 2056–2057, 1996.

[18] X. Li and L. J. Cimini, Jr., "Effect of clipping and filtering on the performance of OFDM," *IEEE Communications Letters*, vol. 2, no. 5, pp. 131–33, 1998.

[19] S. Nobilet, J. F. Herald, and D. Mottier, "Spreading sequences for uplink and downlink MC-CDMA systems: PAPR and MAI minimization," *European Transactions on Telecommunications*, vol. 13, no. 5, pp. 465–474, 2002.

[20] B. A. Ranjha and M. Kavehrad, "Precoding techniques for PAPR reduction in asymmetrically clipped OFDM Based optical wireless systems," in SPIE Photonics West, San Francisco, CA, February 2–7, 2013.

[21] J. R. Barry, J. M. Kahn, W. J. Krause, E. A. Lee, and D. G. Messerschmitt, "Simulation of multipath impulse response for indoor wireless optical channels," *IEEE Journal on Selected Area in Communication*, vol. 11, no. 3, pp. 367–379, 1993.

[22] F. J. Lopez-Hernandez, R. Perez-Jimenez, and A. Santamaria, "Modified Monte Carlo scheme for high-efficiency simulation of the impulse response on diffuse IR wireless indoor channels," *Electronics Letters*, vol. 34, pp. 1819–1820, 1998.

[23] H. Sari, G. Karam, and I. Jeanclaude, "Transmission techniques for digital terrestrial TV broadcasting," *IEEE Communications Magazine*, vol. 33, no. 2, pp. 100–109, 1995.

[24] K. Acolatse, Y. Bar-Ness, and S. K. Wilson, "SCFDE with space-time coding for IM/DD optical wireless communication," in IEEE Wireless Communications and Networking Conference, 2011, Cancun, Mexico, March, 2011.

[25] J. G. Proakis, *Digital Communications*, McGraw Hill, New York, 1995.

[26] Y. P. Lin and S. M. Phong, "BER minimized OFDM systems with channel independent precoders," *IEEE Transaction on Signal Processing*, vol. 51, no. 9, pp. 2369–2380, 2003.

[27] B. Ranjha, Z. Zhou, and M. Kavehrad, "Performance analysis of precoding-based asymmetrically clipped optical orthogonal frequency division multiplexing wireless system in additive white Gaussian noise and indoor multipath channel," *Optical Engineering*, vol. 53, no. 8, p. 086102, 2014.

7

MIMO Technology for Optical Wireless Communications using LED Arrays and Fly-Eye Receivers

7.1 Introduction

Multiple-input and multiple-output (MIMO) is a method of applying multiple antennas at both the transmitter and the receiver to improve communication performance. MIMO is one of the smart antenna technologies. The input and output here indicate communication channels rather than transmitting and receiving devices. Consequently, in indoor optical wireless communications (OWC), MIMO systems refer to the systems with multiple optical channels between the source and the receiver.

People have paid considerable attention to MIMO technology, since it significantly improves data throughput and link range without increasing bandwidth or transmission power. It spreads the power over multiple antennas to improve bandwidth efficiency and/or achieve diversity gain. Due to the advantages of this technique, it has been extensively applied in modern wireless standards, such as IEEE 802.11n (Wi-Fi), LTE, WiMAX, and HSPA+.

7.2 MIMO Configurations

7.2.1 MIMO System Model

The difference between a MIMO and non-MIMO OWC system is that the MIMO system consists of multiple transmitters and receivers. Assuming there are N_t light sources on the ceiling and N_r photo detectors at the receiving device, the received signal vector is given by [1]:

$$\mathbf{y}(t) = \mathbf{H}(t) \otimes \mathbf{s}(t) + \mathbf{n}(t) \tag{7.1}$$

where $\mathbf{y}(t)$ is the received vector, $\mathbf{H}(t)$ is the channel characteristics matrix, $\otimes$ stands for convolution, $\mathbf{s}(t)$ is the transmitted signal vector as $\mathbf{s}(t) = \left[s_1(t), s_2(t), \ldots, s_{N_t}(t) \right]^T$, where $[.]^T$ is

Short-Range Optical Wireless: Theory and Applications, First Edition. Mohsen Kavehrad,
M. I. Sakib Chowdhury and Zhou Zhou.
© 2016 John Wiley & Sons, Ltd. Published 2016 by John Wiley & Sons, Ltd.

the transpose operator, and $\mathbf{n}(t)$ represents the total noise including ambient shot light noise and thermal noise. We assume the noise $\mathbf{n}(t)$ is independent of the transmitted signal and is a real Gaussian random process. It has zero mean with a variance $\sigma^2(t) = \sigma_{shot}^2(t) + \sigma_{thermal}^2(t)$, where $\sigma_{shot}^2(t)$ and $\sigma_{thermal}^2(t)$ are the variances of shot noise and thermal noise, respectively. In practice, multiple sources are generally implemented by LED arrays. As the LED sources are in close proximity to each other and can be driven by the same driver, we suppose they are perfectly synchronized.

7.2.2 Spatial Diversity

Spatial diversity, also called repetition coding (RC), is the most basic application of MIMO technology [2]. Diversity is a very effective remedy in the sense that it exploits the principle of providing the receiver with multiple distorted replicas of the same information bearing signal. Spatial diversity has been extensively studied in conventional wireless systems. Figure 7.1 demonstrates the receiver diversity system. Based on the combining methods, it can be divided into four categories: Selective Combining (SC), Maximal Ratio Combining (MRC), Equal Gain Combining (EGC), and Switch Combining (SSC).

7.2.2.1 Selective Combining

In SC, the branch with the highest signal-to-noise ratio (SNR) is always selected. This can be represented as

$$\tilde{r} = \max\left(\tilde{r}_k\right) \tag{7.2}$$

In a continuous transmission system, SC is not practical because of its requirement of constant monitoring of all the branches. With the availability of this monitoring, the MRC should be a better choice for its better performance compared to SC.

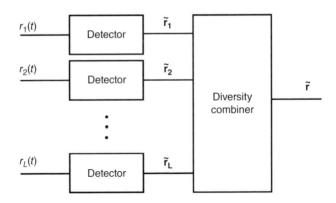

Figure 7.1   A typical receiver diversity system.

7.2.2.2 Maximal Ratio Combining

In a typical MRC diversity system, branches are weighted by their respective complex fading gains and added together. MRC realizes a maximum-likelihood (ML) receiver.

7.2.2.3 Equal Gain Combining

EGC and MRC are similar in that their diversity branches are co-phased. The difference between the two is that EGC does not weight its diversity branches. For a practical consideration, EGC is useful for modulation schemes with equal energy symbols, such as M-phase-shift keying.

7.2.2.4 Switched Combining

The switched combining receivers scan through all the diversity branches until one which reaches a specific SNR threshold is found. This branch is chosen and used until its SNR drops below the threshold. After this, the receiver will rescan all the branches and choose the next one, whose SNR exceeds the threshold. The most conspicuous advantage of SSC is that only one detector is needed in this approach.

7.3 Angle-Diversity Receivers

7.3.1 Angle-Diversity Receiver Overview

In indoor OWC systems, substantial performance advantage can be obtained through applying an angle-diversity receiver. It consists of multiple branches covering different directions [3–6]. Each branch is followed by its own photo detector and amplifier that convert the received optical power of this branch into amplified electrical current for further signal processing. Angle-diversity receivers provide many advantages: first, they reduce ambient light by pointing directly to the desired light source. Also, they reduce ISI and multipath distortions, as most unwanted source interferences and diffusions are rejected from the branch. Finally, the receivers provide wide total field-of-view (FOV).

In Ref. [6], a typical angle-diversity receiver is implemented using multiple non-imaging elements oriented to different directions. The authors systematically studied the theoretical gain of this kind of receiver and demonstrated an OWC system achieving a data rate of 70 Mbps. Figure 7.2 depicts the setup of an angle diversity receiver. Multiple narrow FOV branches collaboratively cover a wide FOV. Ambient light and diffused light that do not coincide with the signal light are blocked out.

7.3.2 Fly-Eye Receiver Design

Yun and Kavehrad proposed several designs of fly-eye receiver [4]. Alignment and focusing are major challenges to fly-eye receiver design [7]. A fly-eye receiver includes multiple independent branches pointing at different directions. Each branch, as a result, is expected to be aligned independently to achieve optimal MIMO gain. In addition, when users roam around

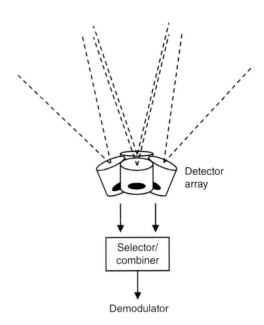

Figure 7.2   A typical configuration of an angle-diversity receiver.

in a typical indoor environment, the distance from a light source to the corresponding receiving branch varies from a few meters to tens of meters.

The general solution to adjust focusing is changing the relative position between the receiver lens and photo detectors. This method, however, requires much adaptive mechanisms and will make the receiver costly and bulky. The authors of Ref. [7] attempted to fix the detectors at one focal length behind the lens and compared its performance with adaptive configurations. It is found that significant simplicity is obtained from the absence of adaptive focusing mechanisms which results from the constant response at dynamic range and the performance similarity of two kinds of designs at long distance.

Fly-eye receivers are expected to view various directions simultaneously. The most straightforward design of fly-eye receivers is employing distributed eyes with their own lens and photo detectors. The large number of lens in this method, however, makes the system complex and bulky. A few novel ideas are proposed to solve the problem.

One instance is applying a transparent ball, glass, or plastic, as the lens for all branches. Because of the symmetry of the ball, all branches share the same ball as demonstrated in Figure 7.3. This design effectively reduces the complexity of the system. A problem of this design is the ball can be very heavy when large aperture is required. This system is most suitable for networks covering small rooms.

For the scenarios where large apertures are required, an alternative approach is utilizing the off-axis imaging ability of a lens. The fly-eye design is depicted in Figure 7.4. This system provides optimal performance for large apertures, especially when a Fresnel lens is applied. The major drawback, when compared with the ball lens, is that different light paths cannot be located far from the optical axis of the lens, as the aberration will be significant. Although increasing photo detector area may improve system tolerance to aberration, the flexibility of this design is limited.

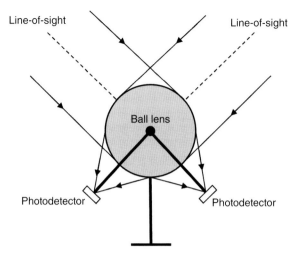

Figure 7.3   Fly-eye design with ball lens.

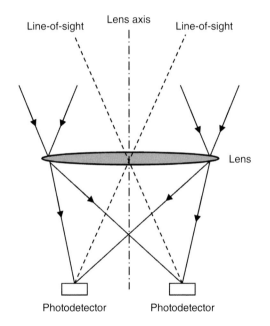

Figure 7.4   Fly-eye design using off-axis imaging.

7.4   Simulation Results and Discussions

7.4.1   Simulation Parameters

The communication environment is assumed to be an ordinary office room. Room dimensions and the reflectance of each surface are reasonably chosen by referring to common conventions. The source array consists of seven sources. They form a hexagon on the ceiling with one

Table 7.1 Simulation parameters of the MIMO OWC system—source array

Number of sources	7
Source location (m)	Source 1: (1.2379 3.0000 3.0000)
	Source 2: (2.1190 1.4740 3.0000)
	Source 3: (3.8810 1.4740 3.0000)
	Source 4: (4.7621 3.0000 3.0000)
	Source 5: (3.8810 4.5260 3.0000)
	Source 6: (2.1190 4.5261 3.0000)
	Source 7: (3.0000 3.0000 3.0000)
Source direction $\left(\hat{\mathbf{i}},\hat{\mathbf{j}},\hat{\mathbf{k}}\right)$	(0 0 −1) for each source
Transmitting power	1 W for each source
Source half-power angle	60° for each source

in the center and six around. The fly-eye receiver consists of seven branches. One branch is placed in the center facing vertically upward and the other six are evenly distributed around the center one with 30° tilt. The fly-eye receiver is located at 841 locations, covering the whole room, for performance test. More details of the source, the fly-eye receiver, environment, and communication parameters are given in Tables 7.1, 7.2, 7.3, and 7.4, respectively.

A single branch non-MIMO receiver model is created for reference. In order to make a fair comparison between MIMO and non-MIMO systems, its aperture area equals the total area of all MIMO receiver branches, and its FOV equals the total FOV of the MIMO receiver. All simulations on the MIMO receivers are repeated on the non-MIMO receiver.

7.4.2 BER Spatial Distributions for MIMO OWC Systems

The BER distributions of the non-MIMO system, the MIMO EGC system, and the MIMO MRC system of 600 Mbps data rate and −105 dBmW noise are given in Figures 7.5, 7.6, 7.7, 7.8, 7.9, and 7.10. The noise level is a reasonable estimation for a regular office environment [8]. The results show that the MIMO systems provide very low outage probabilities. They indicate that the MIMO methods guarantee most areas in the room to have satisfactory communication performance.

In order to make a clearer observation of the BER spatial distributions, we increase the noise level to −90 dBmW, which represents a tough indoor OWC environment involving direct sunlight exposure. The results are shown in Figures 7.11, 7.12, 7.13, 7.14, 7.15, and 7.16.

The following observations can be made from these figures. Generally, the BER distributions are closely related to the source array's layout. We name the source in the array center as the center source and the other six around it as the peripheral sources. The areas in peripheral sources' footprints have low BER. That is because these areas have line-of-sight (LOS) transmission paths with small emitting angles and receiving angles. These factors result in high SNR. Regarding the areas in the footprint of the center source, the situation is more complex. On one hand, in the non-MIMO system, though these areas receive strong signal power from LOS paths from the center source, they also receive LOS signal power from the peripheral sources. Since these LOS signals arrive at different times, they cause inter-source interference.

Table 7.2   Simulation parameters of the MIMO OWC system—receiver

Number of receiver branches	7
Branch direction $(\hat{\mathbf{i}}, \hat{\mathbf{j}}, \hat{\mathbf{k}})$	Branch 1: (0.0000 0.0000 1.0000)
	Branch 2: (0.6428 0.0000 0.7660)
	Branch 3: (0.3214 0.5567 0.7660)
	Branch 4: (−0.3214 0.5567 0.7660)
	Branch 5: (−0.6428 0.0000 0.7660)
	Branch 6: (−0.3214 −0.5567 0.7660)
	Branch 7: (0.3214 −0.5567 0.7660)
Branch responsivity	1 A/W for each branch
Branch field-of-view	20° for each branch
Branch aperture area	1×10^{-4} m² for each branch

Table 7.3   Simulation parameters of the MIMO OWC system—environment

Room dimensions	Length: 6 m
	Width: 6 m
	Height: 3 m
Room surface reflectance	Ceiling: 0.9
	Wall$_{East}$: 0.7
	Wall$_{West}$: 0.7
	Wall$_{North}$: 0.7
	Wall$_{South}$: 0.7
	Floor: 0.1
Spatial resolution	0.02 m
Number of reflections traced	3

Table 7.4   Simulation parameters of the MIMO OWC system—communication system

Test data length	1×10^6
Test data rate	600 Mbps
Test noise level	−120 to −70 dBmW
Test noise increment	1 dBmW

The inter-source interference produces ISI in the communication system, which significantly increases BER. As a result, the areas in the center source footprint in the non-MIMO system have high BER. On the other hand, in MIMO EGC and MIMO MRC systems, the areas in the center source footprint also have LOS links from the center source and the peripheral sources. The MIMO angle diversity mechanism, however, distinguishes the signal power from different sources. By applying different combining algorithms, the ISI is significantly reduced. The areas in the center source footprint, therefore, have low BER in the MIMO systems. For the areas outside the sources' footprints, they have high BER in both MIMO and non-MIMO systems due to low SNR.

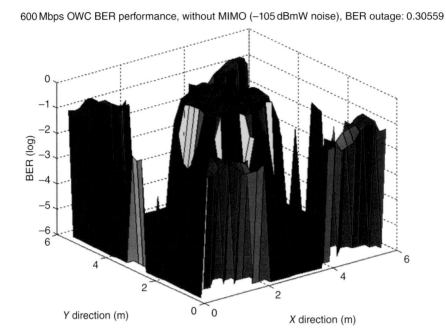

Figure 7.5   Non-MIMO system BER performance of 600 Mbps data rate, −105 dBmW noise.

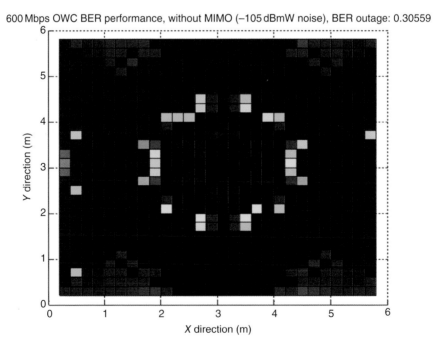

Figure 7.6   Non-MIMO system BER performance of 600 Mbps data rate, −105 dBmW noise (top view).

600 Mbps OWC BER performance, 7 × 7 MIMO using EGC (−105 dBmW noise), BER outage: 0.04042

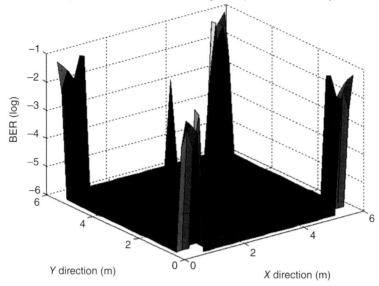

Figure 7.7   MIMO EGC system BER performance of 600 Mbps data rate, −105 dBmW noise.

600 Mbps OWC BER performance, 7 × 7 MIMO using EGC (−105 dBmW noise), BER outage: 0.04042

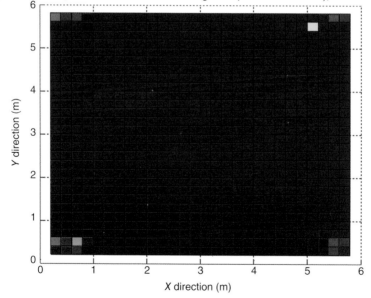

Figure 7.8   MIMO EGC system BER performance of 600 Mbps data rate, −105 dBmW noise (top view).

600 Mbps OWC BER performance, 7×7 MIMO using MRC (−105 dBmW noise), BER outage: 0.00832

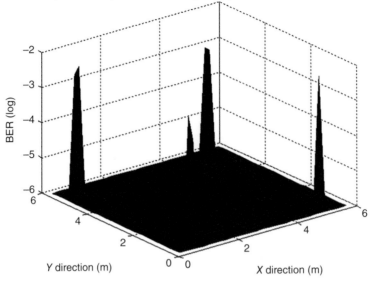

Figure 7.9 MIMO MRC system BER performance of 600 Mbps data rate, −105 dBmW noise.

600 Mbps OWC BER performance, 7×7 MIMO using MRC (−105 dBmW noise), BER outage: 0.00832

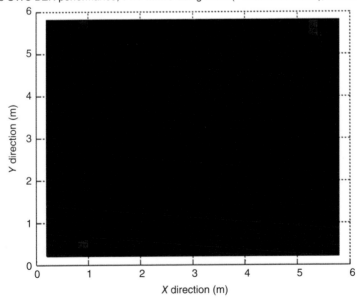

Figure 7.10 MIMO EGC system BER performance of 600 Mbps data rate, −105 dBmW noise (top view).

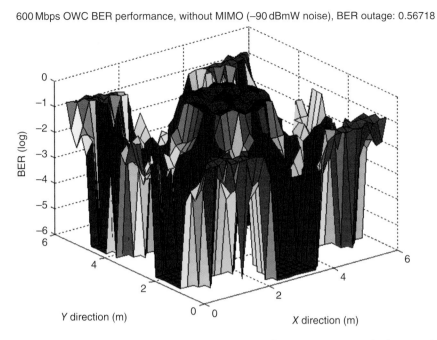

Figure 7.11 Non-MIMO system BER performance of 600 Mbps data rate, −90 dBmW noise.

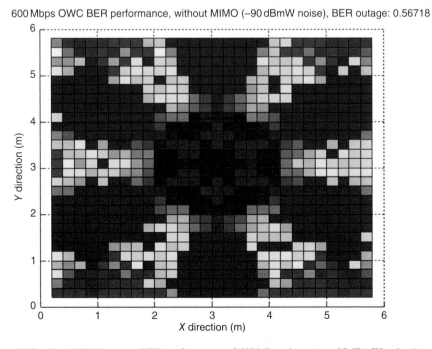

Figure 7.12 Non-MIMO system BER performance of 600 Mbps data rate, −90 dBmW noise (top view).

600 Mbps OWC BER performance, 7×7 MIMO using EGC (−90 dBmW noise), BER outage: 0.46611

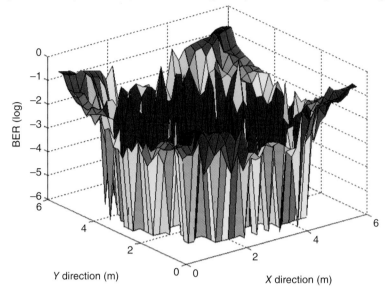

Figure 7.13   MIMO EGC system BER performance of 600 Mbps data rate, −90 dBmW noise.

600 Mbps OWC BER performance, 7×7 MIMO using EGC (−90 dBmW noise), BER outage: 0.46611

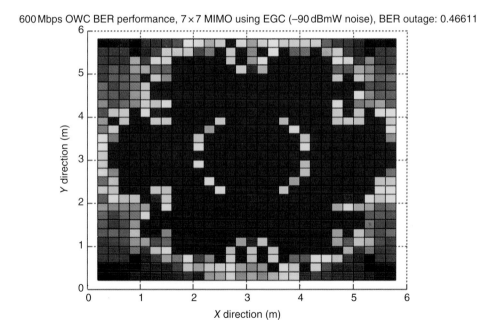

Figure 7.14   MIMO EGC system BER performance of 600 Mbps data rate, −90 dBmW noise (top view).

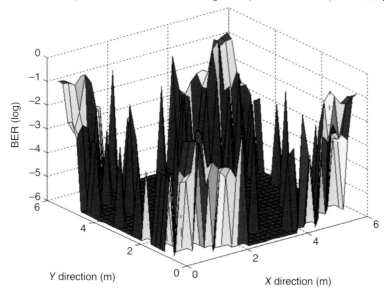

600 Mbps OWC BER performance, 7×7 MIMO using MRC (−90 dBmW noise), BER outage: 0.17241

Figure 7.15 MIMO MRC system BER performance of 600 Mbps data rate, −90 dBmW noise.

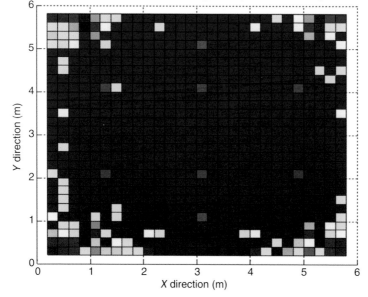

600 Mbps OWC BER performance, 7×7 MIMO using MRC (−90 dBmW noise), BER outage: 0.17241

Figure 7.16 MIMO MRC system BER performance of 600 Mbps data rate, −90 dBmW noise (top view).

7.4.3   Impact of Ambient Noise

In order to explore the impact of ambient noise on system performance, we observe the BER distributions with noise from −75 to −114 dBmW with 600 Mbps data rate. The results for the non-MIMO system, the MIMO EGC system, and the MIMO MRC system are shown in Figures 7.17, 7.18, 7.19, 7.20, 7.21, 7.22, 7.23, 7.24, and 7.25. In these figures, the blue elements indicate that the corresponding areas have low BER and the red ones indicate that the corresponding areas have high BER.

Figures 7.17, 7.18, and 7.19 give the BER distributions of the non-MIMO system with noise decreasing from −70 to −120 dBmW. The low BER areas first appear in the footprints of the peripheral sources. As the noise decreases, these BER areas extend; therefore, the system outage probability reduces. It converges to 32% when the noise level reaches −108 dBmW. An interesting observation from the figures is that the areas in the center source footprint, which are located in the center of the room, never have low BER. That is because the inter-source interference from neighbor sources increases channel delay spread, thus increasing BER.

Figures 7.20, 7.21, and 7.22 give the BER distribution of the MIMO EGC system with noise from −70 to −120 dBmW. Being different from the non-MIMO system, the MIMO EGC

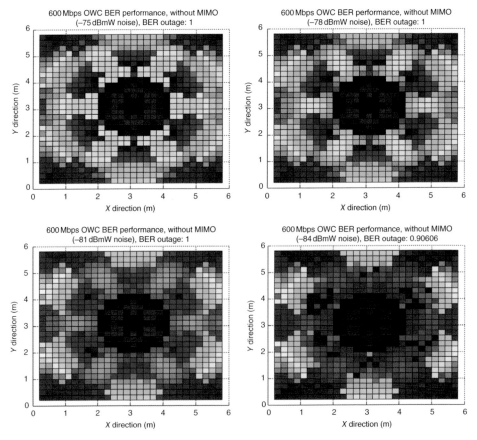

Figure 7.17   BER distributions and outage probabilities of the non-MIMO system (noise from −75 to −84 dBmW).

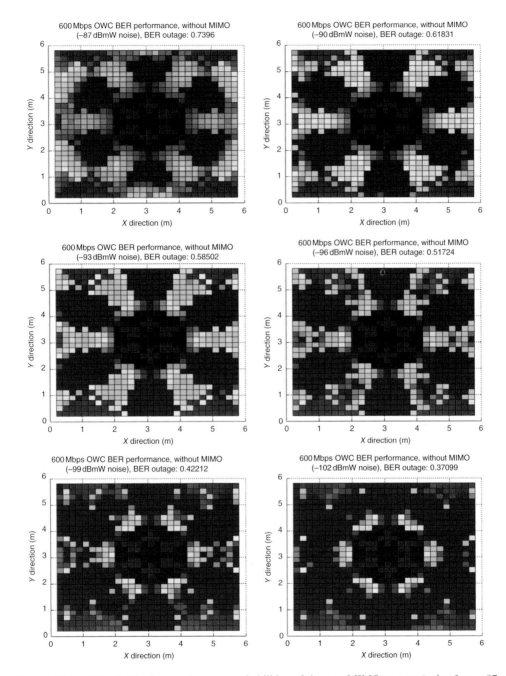

Figure 7.18 BER distributions and outage probabilities of the non-MIMO system (noise from −87 to −102 dBmW).

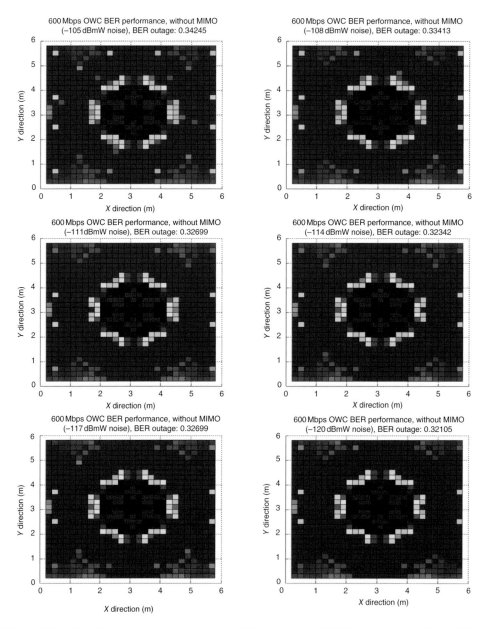

Figure 7.19 BER distributions and outage probabilities of the non-MIMO system (noise from −105 to −120 dBmW).

system has low BER areas first appearing in the center source footprint. As the center low BER areas extend with noise decrease, other low BER areas emerge under the peripheral sources' footprints. They extend with the center, low BER areas as noise decreases. The outage probability converges to 0 at −114 dBmW noise level. An interesting observation is that there are several high BER "islands" surrounded by low BER areas. The reason is that in these locations, multiple sources send light into the most efficient receiver branch through LOS paths and produce inter-source interference.

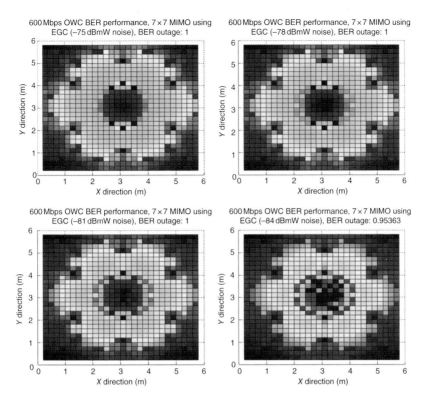

Figure 7.20 BER distributions and outage probabilities of the MIMO EGC system (noise from −75 to −84 dBmW).

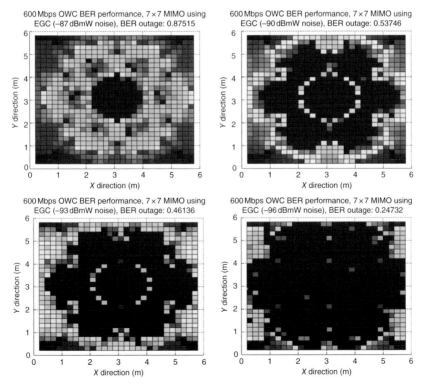

Figure 7.21 BER distributions and outage probabilities of the MIMO EGC system (noise from −87 to −102 dBmW).

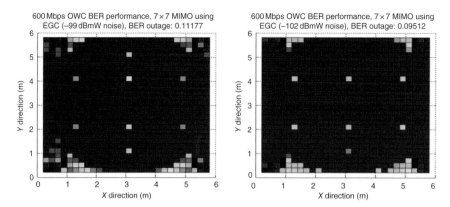

Figure 7.21 (*Continued*)

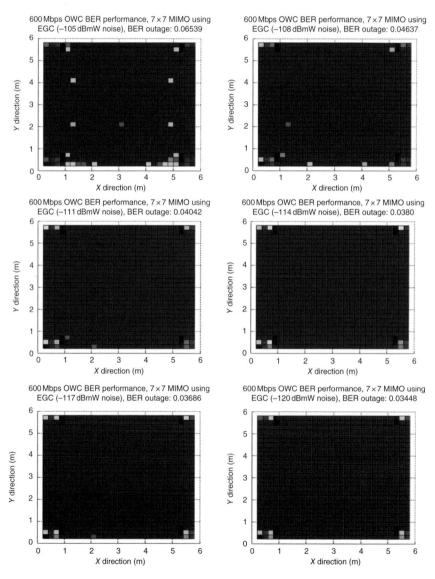

Figure 7.22 BER distributions and outage probabilities of the MIMO EGC system (noise from −105 to −120 dBmW).

The BER performance of the MIMO MRC system is given in Figures 7.23, 7.24, and 7.25. The BER distribution changes in the same pattern as the MIMO EGC system; however, the outage probability converges much faster. It decreases to 0 at the noise level of −102 dBmW. That is because MIMO MRC receivers weight different branches according to their SNRs. High SNR paths are emphasized in the combining. The MIMO MRC, therefore, has a better BER performance than MIMO EGC at the same noise level.

Figure 7.26 demonstrates the outage probability change with noise. An important observation from the results is that the gain of applying MIMO significantly varies with noise. With noise above −84 dBmW, the outage probability of the non-MIMO system is the lowest. That is because in high noise environments, SNR is the dominant factor determining BER. The non-MIMO system has the highest SNR, for it has the largest aperture area and FOV. Though the non-MIMO system suffers more severe ISI than the MIMO systems, its overall performance exceeds the MIMO systems in high noise environments. With noise decrease, the ISI replaces SNR as the dominant factor of system performance. The MIMO EGC system and MIMO MRC system outperform the non-MIMO system at the noise level of −84 and −89 dBmW,

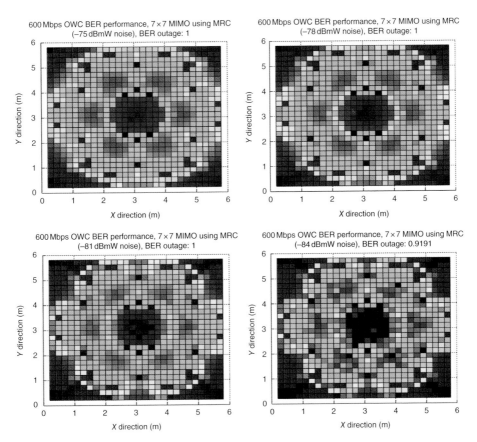

Figure 7.23 BER distributions and outage probabilities of the MIMO MRC system (noise from −75 to −84 dBmW).

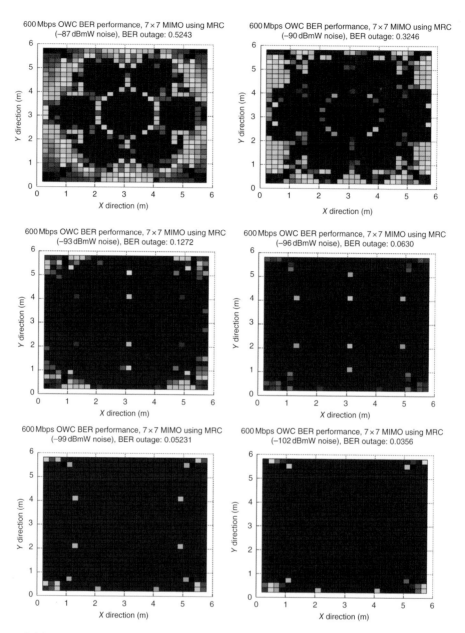

Figure 7.24 BER distributions and outage probabilities of the MIMO MRC system (noise from −87 to −102 dBmW).

respectively. After −89 dBmW, the MIMO systems provide lower outage probability than the non-MIMO system, and the MRC method is always better than the EGC method. The outage probabilities of both MIMO systems converge to zero. The outage probability of the non-MIMO system, however, does not converge to zero. That is because severe ISI remains in the footprint of the center source.

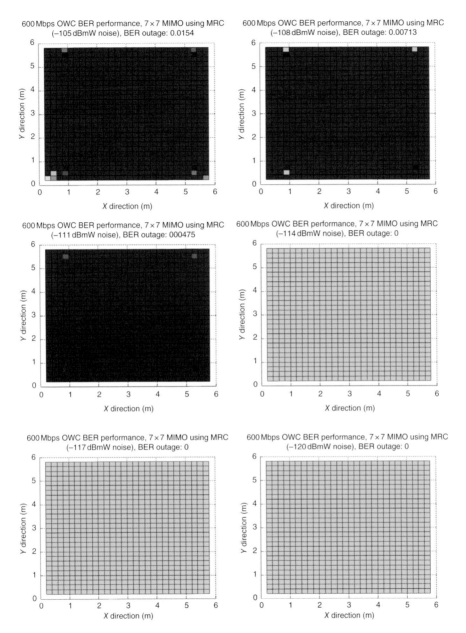

Figure 7.25 BER distributions and outage probabilities of the MIMO MRC system (noise from −105 to −120 dBmW).

7.5   Conclusions

This chapter investigates the application of MIMO technology for indoor OWC. MIMO approaches can improve indoor OWC performance through two mechanisms: one is sending the same data on all MIMO channels to improve communication reliability and we call this

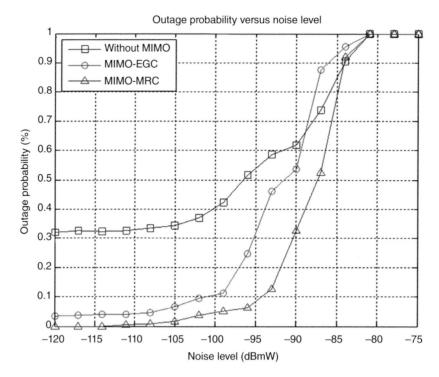

Figure 7.26   Outage probability versus noise level.

method diversity and the other is sending different data on different MIMO channels to increase channel capacity and this method is called multiplexing.

In this chapter, we simulated the BER performance of the MIMO EGC and the MIMO MRC spatial diversity systems, and compared them with the non-MIMO system. The performance is evaluated by BER outage probability. The results indicate that in regular indoor environments, MIMO systems outperform non-MIMO systems. MIMO MRC systems are more efficient than MIMO EGC systems. The reason is that multiple small FOV and aperture branches of MIMO receivers separate the light from different sources, thus mitigating ISI. Nevertheless, when the noise level is high, non-MIMO systems may outperform MIMO systems because non-MIMO systems have a larger aperture size, which become the dominant factor to system performance.

References

[1] T. Fath and H. Haas, "Performance comparison of MIMO techniques for optical wireless communications in indoor environments," *IEEE Transactions on Communications*, vol. 61, no. 2, pp. 733–742, 2012.

[2] G. L. Stüber, *Principles of Mobile Communication*. 2nd ed. Springer, New York, 2011.

[3] J. M. Kahn, R. You, P. Djahani, A. G. Weisbin, B. K. Teik, and A. Tang, "Imaging diversity receivers for high-speed infrared wireless communication," *IEEE Communications Magazine*, vol. 36, no. 12, pp. 88–94, 1998.

[4] G. Yun and M. Kavehrad, "Spot diffusing and fly-eye receivers for indoor infrared wireless communications," in Proceedings of the IEEE Conference on Selected Topics in Wireless Communication, Canada, June 25–26, 1992.

[5] A. M. R. Tavares, R. J. M. T. Valadas, and A. M. de Olveira Duarte, "Performance of an optical sectored receiver for indoor wireless communication systems in presence of artificial and natural noise sources," in Proceedings of the SPIE Conference on Wireless Data Transmission, 1995, vol. 2601, Philadelphia, PA, October 23–25, 1995.

[6] J. B. Carruthers and J. M. Kahn, "Angle diversity for nondirected wireless infrared communications," *IEEE Transactions on Communications*, vol. 48, no. 6, pp. 960–969, 2000.

[7] G. Yun and M. Kavehrad, "Indoor infrared wireless communications using spot diffusing and fly-eye receivers," *Canadian Journal of Electrical and Computer Engineering*, vol. 18, pp. 151–157, 1993.

[8] M. Wolf and D. Kress, "Short-range wireless infrared transmission: the link budget compared to RF," *IEEE Wireless Communications*, vol. 10, no. 2, pp. 8–14, 2003.

8

Wireless Solutions for Aircrafts Based on Optical Wireless Communications and Power Line Communications

8.1 Introduction

To implement wireless communications in aircrafts, the most important concern is the interference of wireless devices to aircraft navigation and communication systems [1]. Personal electronic devices (PED) generate two kinds of microwave radiations: intentional and spurious. Intentional radiations usually are emitted from PEDs having wireless communication functions. They are generated when the devices transmit data via radio frequency (RF) wireless links. Spurious radiations are unintentional but increase the RF noise level. Intentional radiations are not commonly considered as interference because of the strict band limitations as given in Table 8.1 [2]. The accumulated spurious radiations, however, can be very high in a frequency for aircraft navigation and communication systems; thus they cause considerable impact on operations. Studies point out that laptop computers are most frequently suspected as sources of interference. In 40 PED-related reports collected by the International Air Transport Association (IATA), laptop computers are seen to contribute 40% of them. The IATA reports also indicate that the navigation system is the most frequently affected system. According to the reports, 68% of the cases happened in the navigation system [3]. Another study on Aviation Safety Reporting System (ASRS) database confirmed that the navigation system is most vulnerable to PED interference where cell phones and laptops are the most common causes [4].

Recently, researchers have shown considerable interest in using optical wireless communications (OWC) in airplanes and space vehicles [5]. Light-emitting diode (LED) is widely applied for illumination in new generation commercial aircrafts. OWC demonstrates significant advantages in size, weight, power, cost, and electromagnetic interference (EMI) reduction. In

Short-Range Optical Wireless: Theory and Applications, First Edition. Mohsen Kavehrad,
M. I. Sakib Chowdhury and Zhou Zhou.

Table 8.1 Frequency separation (aircraft navigation and communication are shaded)

Omega navigation	ADF	HF	Marker beacon	VOR, localizer	VHF COM	Glide slope
10–14 kHz	190–1750 kHz	2–30 MHz	74.85, 75, 75.15 MHz	108–118 MHz	118–136 MHz	328–335 MHz
GSM 400 450.4–467.6 MHz 478.8–496 MHz	GSM 850 824–894 MHz	GSM 900 876–960 MHz	DME 960–1220 MHz	TCAS/ATC 1030, 1090 MHz	GPS 1575 MHz	SATCOM 1529, 1661 MHz
GSM 1800	European UMTS	GSM 1900	IMS band: WLAN802.11b,g, Bluetooth, Home RF	Low-range altimeter	Microwave landing system	WLAN 802.11a
1710–1880 MHz	1880–2025 MHz 2110–2200 MHz	1850–1900 MHz	2446.5–2483.5 MHz	4.3 GHz	5.03, 5.09 GHz	5150–5350 MHz
Weather radar 5.4 GHz	WLAN 802.11a 5725–5825 MHz	Weather radar 9.3 GHz	Sky radio 11.7 GHz	DBS TV 12.2–12.7 GHz	Frequency separation regulated by International Telecommunication Union, Geneva, Switzerland	

addition, OWC combined with powerline communications (PLC) creates an efficient delivery mechanism for fulfilling broadband access onboard an aircraft, while providing efficient and economic lighting [6]. This chapter explores the potential capabilities of these two emerging techniques.

8.2 Powerline Communications Channel Model

In an aircraft powerline grid, signal propagation does not take place along a direct path from a transmitter to a receiver. Echoes exist because of the reflections at grid junctions and they cause multipath distortion. At the receiver, each transmission path is weighted by a coefficient g, which is defined as the product of the reflection coefficient and the transmission coefficient of the nodes along the transmission path. Since both reflection and transmission coefficients are not greater than one, the weighting coefficient g is always equal to or smaller than unity. By applying these weighting coefficients, the grid channel model can be formulated as the summation of multiple paths with different lengths and weighting factors. As a result, the channel model of a powerline network can be expressed as

$$H(f) = \sum_{i=1}^{N} g_i e^{-\alpha(f)d_i} e^{-j\beta(f)d_i} \tag{8.1}$$

where N indicates the number of significant arrived paths at the receiver, d_i is the length of the ith path, and g_i is the weighting factor of the ith path.

Based on this model, the theoretical Shannon capacity of an aircraft PLC network can be calculated. The simulation result is shown in Figure 8.1 [6]. It indicates that 326 Mbps throughput can be achieved at a transmission power of 20 dBm ($N_0 = -117$ dBm Hz^{-1} is assumed).

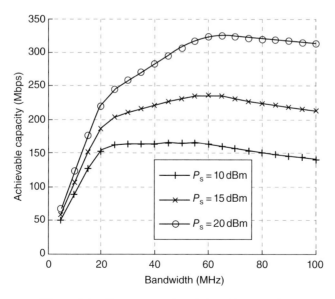

Figure 8.1   Shannon capacity of aircraft's powerline.

8.3   Optical Wireless Communications

8.3.1   Simulation Configurations

Using OWC simulation methods described in previous chapters, we can analyze OWC performances in most environments by carefully customizing environment parameters. In this section,
we explore visible light propagation features and OWC performance in airplane cabins. Based
on the findings, the possibility of applying LED Visible Light Communications (VLC) in airplane cabins for high-speed wireless transmissions is validated. The description here focuses on
the received power and delay-spread distribution in the cabin when using existing overhead LED
reading lights as VLC transmitters. We will simulate the environment of one passenger seating
row in a typical Boeing 737-900 airplane. The interior dimensions are shown in Figure 8.2.

Different surfaces in the cabin are made from different materials and therefore have different
reflectance values. We can choose material samples for cabin surfaces from Ref. [7] and look
up their corresponding reflectance values. The results are given in Table 8.2. Source and receiver
profiles are given in Tables 8.3 and 8.4, respectively.

By tracing the bounces of lights, we have simulated the impulse response from the light
sources to all locations in the seating area. Received power and delay-spread for each location
are further computed based on individual impulse responses. They are demonstrated as their
spatial distributions and are shown in Figures 8.4 and 8.5.

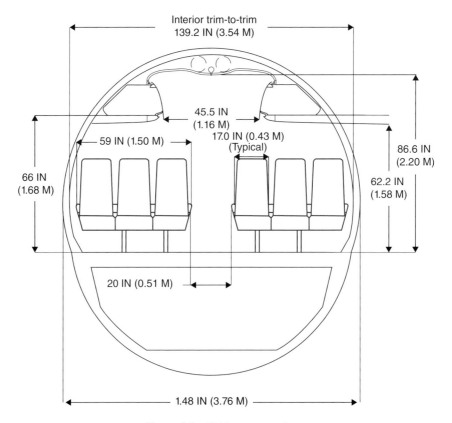

Figure 8.2   Cabin cross-sections.

Table 8.2   Material samples and reflectance for surfaces in the cabin

Surface	Material	Reflectance
Seatback	Rough plastic	0.5
Upholsteries	Linen	0.3
Floor	Rug	0.1
Ceiling	Rough plastic	0.5
Cabin interior wall	Rough plastic	0.5

Table 8.3   Source profiles

Model	2LA455953
Light color	Warm white
Beam angle	15°
Operating current	700 mA max
Power consumption	3.2 W max
Source locations	(10, 7, 45)
	(29, 7, 45)
	(48, 7, 45)

Table 8.4   Receiver profiles

Field-of-view	60°
Responsivity	1 A/W
Aperture area	$1 \times 10^{-4}\,\mathrm{m}^2$
Photo current from background noise	$1 \times 10^{-5}\,\mu\mathrm{A}$

8.3.2   *Illuminance Distribution Results*

In photometry, illuminance, formerly brightness, is the total luminous flux incident on a surface per unit area. The unit of illuminance is lux (lx), or lumen per square meter. Photometric and radiometric parameters are listed in Tables 8.5 and 8.6, respectively.

The photometric parameter luminous flux (unit: lm) and the radiometric parameter radiant power (unit: W) can be calculated by (8.2) and (8.3), respectively:

$$F = 683 \int\limits_{380\ \mathrm{nm}}^{780\ \mathrm{nm}} S(\lambda)V(\lambda)d\lambda \tag{8.2}$$

$$P = 683 \int\limits_{\lambda_L}^{\lambda_H} S(\lambda)d\lambda \tag{8.3}$$

where $S(\lambda)$ is the spectral power and $V(\lambda)$ is the luminous efficiency function, or "photopic" function. The "photopic" function correlates with human brightness perception and is given in Figure 8.3.

Table 8.5   Photometric parameters

Symbol	Quantity	SI unit	Abbreviation
F	Luminous flux	Lumen	lm
I_v	Luminous intensity	Candela (=lm sr^{-1})	cd
E_v	Illuminance	Lux (=lm m^{-2})	lx
M	Luminous efficiency	Lumen per watt	lm W^{-1}

Table 8.6   Radiometric parameters

Symbol	Quantity	SI unit	Abbreviation
P	Radiant power	Watt	W
I	Radiant intensity	Watt per solid angle	W sr^{-1}
E	Irradiance	Watt per square meter	W m^{-2}
L	Radiance	Watt per steradian per square meter	W sr^{-1} m^{-2}

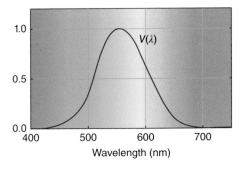

Figure 8.3   Luminous efficiency function.

From the illumination point of consideration, specific illuminance level should be maintained for different applications. From the communication point of consideration, sufficient illuminance should be guaranteed for high signal-to-noise ratio (SNR) [8].

Figure 8.4 is the illuminance distribution of OWC in airplane cabins. The environment model is established by using interior dimensions shown in Figure 8.2 and material samples in Table 8.2. The color of each element in the figure indicates the illuminance level of the corresponding area. For consistency in our results, in all the following figures of this chapter, we use blue for the areas where relatively better communication performance is achieved and red for the areas where relatively worse communication performance is achieved.

We can make the following observations from the figure. First, the center area has the highest illuminance and it decreases in the peripheral areas. Second, optical wireless is able to provide sufficient illuminance for illumination. Illuminance requirement for using computer monitor is 30 lx and for reading it is 300 lx [9]. One hundred percent of seating area satisfies monitor operation requirement, with a minimum illuminance of 44.08 lx and 100% of reading area satisfies reading requirement, with a minimum of 470.27 lx. Third, optical wireless is able to provide sufficient SNR for communications. Ten-decibels SNR is required to provide

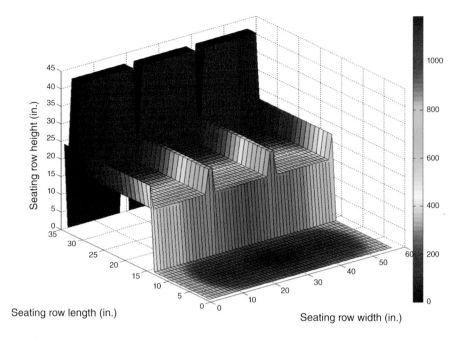

Figure 8.4 Illuminance distribution of LED reading lights in a typical cabin environment on Boeing 737-900.

10^{-5} bit-error-rate (BER) for basic on–off keying (OOK) modulation in additive white Gaussian noise (AWGN) channel [8]. One hundred percent of seating area satisfies the illuminance requirement, with the minimum SNR of 15.11 dB.

8.3.3 Delay Spread Distribution Results

Delay spread is the difference between the time of arrival of the earliest significant multipath component and that of the latest. It is a measure of the multipath richness of a communication channel and indicates the channel bandwidth.

Delay spread is most commonly quantified through root-mean-square (rms) delay spread as

$$s = \sqrt{\frac{\int_{-\infty}^{\infty} (t-\mu)^2 h^2(t)dt}{\int_{-\infty}^{\infty} h^2(t)dt}} \qquad (8.4)$$

where $h(t)$ is the impulse response and μ is average delay defined by

$$\mu = \sqrt{\frac{\int_{-\infty}^{\infty} t h^2(t)dt}{\int_{-\infty}^{\infty} h^2(t)dt}} \qquad (8.5)$$

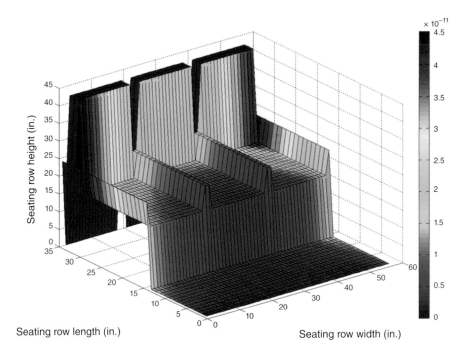

Figure 8.5 Delay-spread distribution of LED VLC system on Boeing 737-900.

Previous research shows that the maximum transmission rate over a wireless channel is one to several times of the inverse of its delay spread, given that no diversity or equalization is applied.

Figure 8.5 is the delay spread distribution in an aircraft VLC system. Like illuminance distribution, the center area has the lowest delay spread and it increases in the peripheral areas. Defining bandwidth as the inverse of delay spread, 100% of the seating area is able to provide hundreds of megahertz bandwidth.

Commercial LEDs provide bandwidth from a few megahertz to tens of megahertz; advanced LEDs are able to provide a few hundred megahertz to gigahertz [10, 11]. Commercial photodiodes have reached hundreds of megahertz [12]. The overall bandwidth of a communication system is jointly determined by source, channel, and receiver. The results here show that channel is not the bandwidth bottleneck in a cabin environment.

8.3.4 Bit-Error-Rate Distribution and Outage Probability

Bit-error-rate (BER) is the number of bit errors divided by the total number of received bits of a data stream over a communication channel. It is the most direct measure of communication performance. A bit error rate of better than 10^{-5} is considered acceptable in wireless local area network (LAN) applications [13]. As BER varies at different locations in the cabin, we apply BER outage probability to evaluate network coverage with satisfactory BER.

BER outage [14] is defined as the fraction of the useful area over which the BER criterion is not satisfied. It is a statistical index of indoor wireless network performance. It indicates the portion of the room area, where the BER is above a criterion, or in other words, lower outage, larger reliable communication area. The flowchart for BER outage calculation is shown in Figure 8.6.

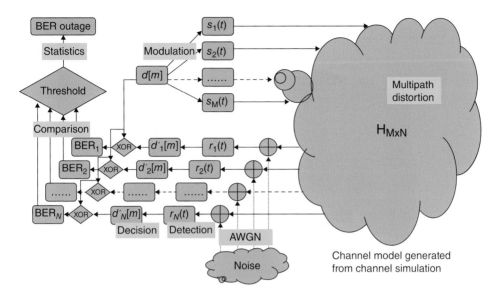

Figure 8.6 BER outage probability calculation flowchart.

BER distribution of in-cabin LED VLC system, 3000 Mbps, BER outage = 0.17261

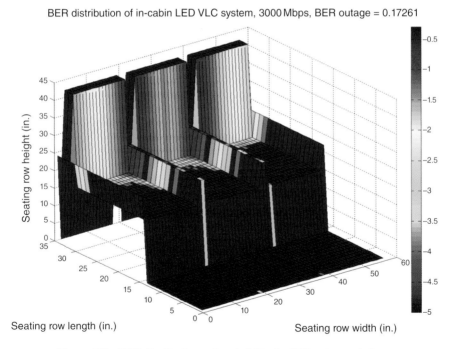

Figure 8.7 BER distribution and probability for 3 Gbps transmission.

BER distribution and outage probability for bitrates from 3 Gbps to 375 Mbps are given in Figures 8.7, 8.8, 8.9 and 8.10.

Top views provide us a better understanding of BER distribution in the cabin. Figure 8.11 is a group of BER top views from 3 Gbps to 23 Mbps.

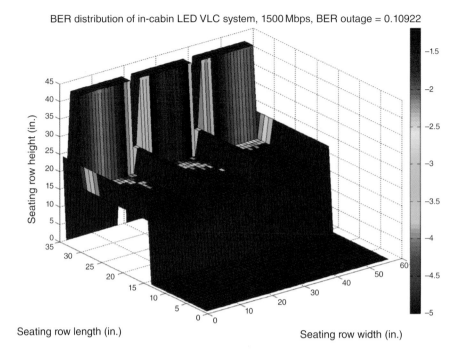

Figure 8.8 BER distribution and probability for 1.5 Gbps transmission.

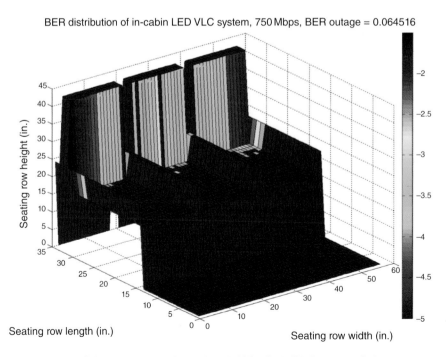

Figure 8.9 BER distribution and probability for 750 Mbps transmission.

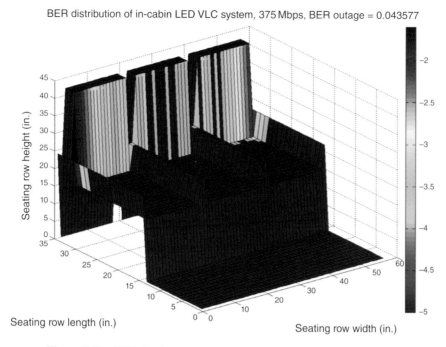

Figure 8.10 BER distribution and probability for 375 Mbps transmission.

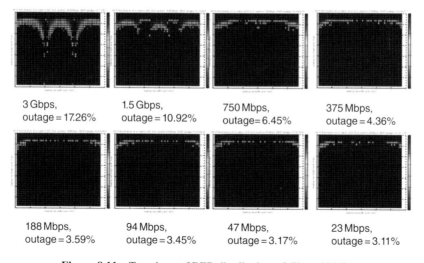

Figure 8.11 Top views of BER distributions, 3 Gbps–23 Mbps.

Figure 8.12 gives the statistical BER outage probabilities for different bitrates. The bitrate requirements for popular multimedia services are provided for reference in Table 8.7.

From these results, we observe that BER outage decreases as bitrate decreases. It reduces to less than 4% at 200 Mbps. Only small high-BER area remains in the peripheral areas due to low SNR. We therefore conclude that optical wireless is able to support high-speed multimedia services in the aircraft cabin environment.

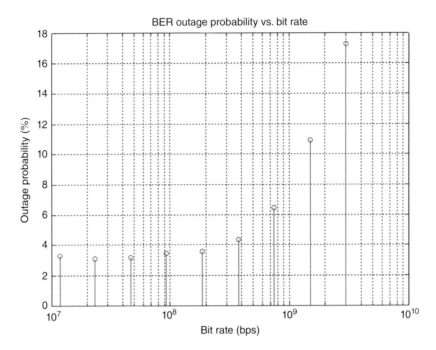

Figure 8.12 BER outage probabilities with bitrates.

Table 8.7 Bitrate requirements for common multimedia services [15]

Service	Bitrate requirement
Blu-ray Disc	40 Mbps
HDV 720p MPEG2	19 Mbps
DVD MPEG2	9.8 Mbps
Mp3 music services	Up to 320 kbps
Phone services	8 kbps
Teleconference services	384 kbps

8.4   Wireless Applications for Commercial Airplanes

Airplane manufacturer Boeing Company discussed several emerging wireless applications in commercial airplanes in Ref. [16]. The PLC–OWC solution is a strong candidate to implement these ideas for its merits in size, weight, power consumption, EMI safety, and labor cost.

8.4.1   Reading Light Passenger Service Units

The existing reading lights above passenger sitting area can provide two services. One is general illumination and the other is data, voice, and video communications.

8.4.2   Passenger Infotainment

Passenger infotainment system is interconnected by visible light between seats to form a high-speed mesh network. It projects to offer in-flight entertainment. Research shows PLC–OWC can provide sufficient bandwidth for this application in a commercial airplane cabin that may have 400 or more passengers.

8.4.3   Cabin Interphones

The lighting LED can transmit low bandwidth voice signal to flight attendants' head phones.

8.4.4   Interconnection of Line-Replaceable-Units Over Environmental Barrier

It is not possible to interconnect the line replaceable units (LRUs) through enclosed areas with fiber. OWC provides optical transmission through these barriers without routing long distance.

8.5   Conclusions

This chapter investigates the PLC-OWC solution for wireless communications in airplane cabins. At first, we discussed the PLC channel model and the estimated capacity. Next, we investigated OWC performance in airplane cabin, when LED is used as transmitter. The cabin environment model is established by the interior dimensions of a typical Boeing 737-900 commercial plane. Using this model, we simulated visible light propagation characteristics and calculated power and delay-spread distributions. In addition, based on the impulse response matrix obtained, we simulated BER distribution and explored BER outage probabilities versus bitrates.

PLC–OWC provides an economical, lightweight, high-speed, energy-saving, and EMI-free solution. The results confirm the possibility of applying this technology for high-speed in-cabin wireless services.

References

[1] O. Orjih, "Recent Developments in Aircraft Wireless Networks," [Available Online]: http://www.cse.wustl.edu/~jain/cse574-06/ftp/aircraft_wireless/ (accessed April 21, 2015).
[2] A. Jahn, M. Holzbock, J. Muller, R. Kebel, M. De Sanctis, A. Rogoyski, E. Trachtman, O. Franzrahe, M. Werner, and F. Hu, "Evolution of aeronautical communications for personal and multimedia services," *IEEE Communications Magazine*, vol. 41, no. 7, pp. 36–43, 2003.
[3] House of Representatives report on the potential hazard of PEDs on aircraft, "Portable Electronic Devices: Do They Really Pose a Safety Hazard on Aircraft," [Available Online]: http://commdocs.house.gov/committees/Trans/hpw106-102.000/hpw106-102_0.HTM (accessed April 21, 2015).
[4] NASA report on cases of PED interference in aircrafts, "Personal Electronic Devices and Their Interference With Aircraft Systems," [Available Online]: http://ntrs.nasa.gov/archive/nasa/casi.ntrs.nasa.gov/20010066904.pdf (accessed April 21, 2015).
[5] M. Kavehrad, "Broadband room service by light," *Scientific American*, vol. 297, no. 1, pp. 82–87, 2007.
[6] M. Kavehrad, Z. Hajjarian, and A. Enteshari, "Energy-efficient broadband data communications using white LEDs on aircraft powerlines," in IEEE Integrated Communications, Navigation and Surveillance Conference, 2008, Bethesda, MD, May, 2008.

[7] K. J. Dana, B. van Ginneken, S. K. Nayar, and J. J. Koenderink, "Reflectance and texture of real-world surfaces," *ACM Transactions on Graphics*, vol. 18, no. 1, p. 1, 1999.

[8] J. G. Proakis and D. G. Manolakis, *Digital Communications*. McGraw-hill, New York, 1995.

[9] *Luminance Level in School Applications* [Available Online]: http://www.e2energysolutions.com/Resources/Light%20levels%20in%20schools%20all%20interior%20and%20exterior%20-%20USE.pdf (accessed April 21, 2015).

[10] J. Cao, Z. Liang, and Z. Ma, "White LED modulation bandwidth and modulation characteristics of the study," *Hans Journal of Wireless Communications*, vol. 2, pp. 7–12, 2012.

[11] R. D. Koudelka and J. M. Woodall, *Light Emitting Devices with Increased Modulation Bandwidth*. Yale University, New Haven, CT, 2011.

[12] *High-Speed Photodetectors*, Thorlabs, Inc. [Available Online]: http://www.thorlabs.us/newgrouppage9.cfm?objectgroup_id=1295 (accessed April 15, 2015).

[13] J. Yee and H. Esfahani, "Understanding wireless LAN performance trade-offs," *EETimes*, November 2002, [Available Online]: http://www.eetimes.com/design/communications-design/4143150/Understanding-Wireless-LAN-Performance-Trade-Offs (accessed April 21, 2015).

[14] M. Kavehrad and P. J. McLane, "Spread spectrum for indoor digital radio," *IEEE Communications Magazine*, vol. 25, no. 16, pp. 32–40, 1987.

[15] Wikipedia, "List of Multimedia Bit Rate," [Available Online]: http://en.wikipedia.org/wiki/Bit_rate#Multimedia (accessed April 21, 2015).

[16] E. Chan, "Wireless optical links for airplane applications," in IEEE Photonics Society Summer Topical Meeting Series, 2012, Washington, DC, July 2012.

9

Multispot Diffusing Transmitters Using Holographic Diffusers for Infrared Beams and Receivers Using Holographic Mirrors

9.1 Introduction

We have discussed about Visible Light Communications (VLC) based on white LEDs as a potential enabling technology for Optical Wireless (OW) systems. LEDs are not very suitable for providing a high data rate though. For high data rate transmissions, usually laser sources are used in the infrared range. Advantages of infrared (IR) wireless local access over its radio frequency (RF) counterpart are the same as general OW systems, that is, a large unregulated band, enhanced scope for spatial reuse of the available bandwidth, absence of electromagnetic interference, reduced detector sensitivity to spatial movements, and availability of low power consumption light sources.

There are two basic classification schemes for wireless IR links. In the first approach, the link can be directed, employing a narrow beam transmitter and a narrow field-of-view (FOV) receiver, or non-directed, with a broad beam transmitter and a wide FOV receiver. The second scheme classifies the links according to whether or not they rely on a line-of-sight (LOS) between the transmitter and the receiver (LOS and non-LOS configurations). The directed LOS infrared method is the most efficient one in power consumption and can achieve very high bit rates. Its drawbacks are tight alignment requirement, immobility of the receiver, and interruptions in transmission caused by shadowing. These disadvantages are overcome in non-directed non-LOS methods (referred to as diffuse links), which utilize diffuse reflections from the ceiling and walls. On the other hand, the latter methods suffer from ineffective use of power and multipath dispersion, which tend to greatly limit the rate of transmission. Cellular architecture (non-directed, LOS) employs a wide FOV receiver and a broad beam transmitter retaining the LOS and combines the mobility of the diffuse and the high-speed capability of the LOS systems. It may still suffer from blockage and shadowing. Another way

Short-Range Optical Wireless: Theory and Applications, First Edition. Mohsen Kavehrad, M. I. Sakib Chowdhury and Zhou Zhou.
© 2016 John Wiley & Sons, Ltd. Published 2016 by John Wiley & Sons, Ltd.

to combine the advantages and to overcome the drawbacks of the directed LOS and diffuse configurations is to use a transmitter producing multiple diffusing spots and angle-diversity detection, proposed by G. Yun and M. Kavehrad [1]. This configuration relies on multiple LOS between the diffusing spots and the receiver. There are some practical means to create multiple diffusing spots. E. Simova *et al.* [2, 3] combined the diffusing and splitting functions in a single holographic optical element (HOE), while J. Carruthers and J. Kahn [4] used eight laser diodes to produce eight collimated beams. Both groups demonstrated an improvement in the overall system performance using a multiple-beam transmitter. Angle diversity detection can be performed in two main ways: using a composite receiver with several branches looking at different directions [4–10] or using an imaging receiver consisting of imaging optics and an array of photo detectors [1, 10–12].

In this chapter, we investigate the use of a computer-generated hologram (CGH) as a beam-splitting element for a multiple-beam transmitter for indoor wireless infrared links. We present simulation results for the channel parameters when such a transmitter is utilized over a communication link. Comparison with a pure diffuse link is made in order to distinguish the characteristic features of the multispot diffusing configuration (MSDC). A communication link utilizing angle diversity detection with a seven-branch composite receiver is computer simulated and is compared to the case when a single element wide FOV receiver is used. The system robustness against shadowing and blockage is also discussed.

9.2   CGH for Intensity-Weighted Spot Arrays

For our application, we would like to have a transmitter that produces multiple beams with prescribed intensities covering the ceiling of a room, which may have an irregular form (not rectangular). There are many ways to produce multiple beams. The most straightforward approach is to have several light sources aiming at different directions. The beams can have desired intensities and an area of any shape can be covered. However, practically, we cannot have a large number of beams.

As a good alternative, HOEs can be used at the transmitter to generate multiple beams from a single laser diode. The holograms can be fabricated by conventional optical means, utilizing a multiple-exposure technique. An exact prescribed ratio between the intensities of the beams cannot be achieved with this technique [2]. This leads to a non-homogeneous distribution of optical power within a room. Furthermore, this technique cannot be used for very large spot arrays and for asymmetrical spot arrays. Alternatively, holograms generated by means of a computer can produce wave-fronts with any prescribed amplitude and phase distribution. CGH, as in Figure 9.1, have many useful properties. An ideal wavefront can be computed on the basis of diffraction theory and encoded into a tangible hologram. A multilevel phase CGH can have a diffraction efficiency close to 100%. HOEs have insignificant physical weight and are economic when mass produced.

For our application, the HOE should be capable of producing multiple beams with prescribed intensities, thus distributing the optical power as uniformly as possible in a given area. If the hologram consists of a great number of periodic replications of a single elementary cell (as in Figure 9.2a) and is illuminated by a collimated laser beam, the far field diffraction pattern is a lattice of spots, the lattice spacing ΔS being determined by the size of the cell L [13, 14] (Figure 9.1): $\Delta S = \lambda F / L$, where λ is the wavelength of illumination, and F is the

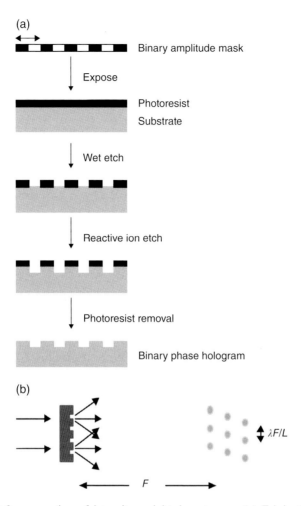

Figure 9.1 CGH for generation of intensity weighted spot array. (a) Fabrication and (b) optical arrangement for far-field pattern observation.

distance between the hologram and the observation plane. Since the area that has to be covered by the spot array is very large, the size of the elementary hologram cell has to be extremely small. The amplitudes and phases of every spot are determined by the elementary cell pattern and are given by its Fourier transform modes. We are not interested in the relative phases of the generated spots. This provides more freedom in the design process and allows higher diffraction efficiencies. To design the required holographic beam-splitter, the hologram elementary cell is broken up into a square array of pixels, each of them imparting a specified phase delay to the incident wavefront. The iterative encoding design method [13] can be employed, in which the hologram cell is built up through gradual selections of changes, pixel-by-pixel, from a random initial cell pattern. The technique of simulated annealing [15] can be used to minimize the cost function, defined from the difference between the desired spot pattern and the actual output pattern. In Figure 9.2a, a bilevel hologram pattern is shown.

(a)

Elementary cell   4 × 4 repeats of the elementary cell

(b)

First mask   Second mask   Third mask

Elementary cell pattern

Figure 9.2 (a) Bilevel CGH producing 8×8 beams with 80% diffraction efficiency; spot intensity variation less than 3.5%, and (b) 8-level CGH producing 10×10 beams with 87% diffraction efficiency; spot intensity variation less than 1.5%.

The hologram is designed to produce an array of 8×8 spots with 80% diffraction efficiency. The intensity variation between the spots is less than 3.5%. An example for an 8-level CGH producing 10×10 beams with 87% diffraction efficiency and beam intensity variation less than 1.5% is shown in Figure 9.2b.

The number N of the phase levels of the hologram determines the number n of the masks that have to be made: $N = 2^n$. The more phase levels, the more complexity in the fabrication of the hologram. The fabrication process of a bilevel hologram is shown in Figure 9.1a. From the binary computer-generated amplitude mask, a surface relief binary phase hologram is produced through etching in a substrate. In the case of a multilevel hologram, this process is repeated for each of the masks. Each etch step is produced with half the depth of the previous etch step. Thus, the combination of the three etchings with the three binary amplitude masks, shown in Figure 9.2b, generates eight phase levels in the final hologram. Though the bilevel hologram is easier for fabrication, it has a lower diffraction efficiency compared to the multi-level holograms, restricts the shape of the spot array to a symmetrical one, and is not capable of producing spots with prescribed but unequal intensities. If the area that has to be illuminated is, for example, L-shaped, a multilevel hologram has to be used.

The choice of the number of spots depends on the detection method. If wide FOV receivers are to be used, uniform illumination of the ceiling would be the best. This can be achieved with the use of a holographic beam-splitter that produces as many beams with equal intensity as possible and is illuminated with a slightly divergent laser beam, causing overlapping of the far-field beam patterns, thus producing a uniformly illuminated area. Such illumination will distribute the optical power uniformly over the ceiling. A similar technique was used by the BT Laboratories group [16, 17] to produce optical cells with different shapes for a cellular IR wireless architecture. If angle diversity detection is used, joint optimization of receiver FOV and the number of diffusing spots will be imperative. In principle, the more the diffusing spots are, the more optical power can be transmitted in compliance with eye safety regulations. In addition, a more uniform distribution and a higher predictability of the channel parameters can be achieved. Since the beams emerging from the hologram are almost collimated, eye safety limits for a point source should be observed for each one of the beams. Theoretically, if the hologram produces $M \times N$ beams with equal intensity, the total optical power that can be safely launched is $M \times N$ times the accessible emission limit for a point source. In practice, the limit for the total optical power depends not only on the total number of emerging beams but also on the power contained in the non-diffracted part of the initial laser beam that forms the so-called central spot. The central spot intensity is determined by the hologram design and the accuracy of the fabrication process [17]. Let us assume that the central spot contains 1% of the laser power, which is practically achievable. Then increasing the number of the diffusing spots up to 100 will allow increasing the intensity of each spot and of the total transmitted optical power. For larger numbers of spots, the total permissible power will remain the same, while the intensity of a single spot will decrease.

9.3 Communication Cells for Multispot Diffusing Configuration

One of the problems with both diffuse and multispot diffuse (quasi-diffuse) systems is the relatively small coverage range. In diffuse configuration, there is a severe increase in the optical path loss as the distance between the receiver and the transmitter increases. Though the latter is not true for quasi-diffuse configurations, still the communication cell size is limited by other factors. In order to cover a large area, the transmitter has to emit beams at very small elevation from a horizontal line (Figure 9.3). First, it makes the system sensitive to shadowing of the transmitter by people moving, which can be avoided by placing the transmitter at a human height. Second, there is a restriction about the angle by which the transmitted beams strike the reflecting surface. Both diffuse and quasi-diffuse links rely on diffuse reflections from reflecting surfaces. Increasing the angle of incidence above 60°, the reflection pattern of typical office surfaces (ceiling and walls) deteriorates from Lambertian reflector and at 70° exhibits strong specular reflections [18]. The natural approach to covering very large rooms is to use several transmitters, thus forming several communication cells. Apparently, the maximum horizontal dimension D of a single communication cell depends on

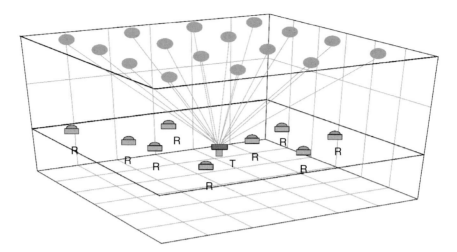

Figure 9.3 Multispot diffusing configuration. R, receiver; T, transmitter.

the distance H_t between the transmitter and the reflecting surface (ceiling): $D \approx 2H_t \tan 65°$. For instance, cell diameters of 8.6 m and 12.9 m correspond to distances between the transmitter and the ceiling of 2 m and 3 m, respectively. In multi-spot diffusing configuration (MSDC), the shape of the cell is chosen to be square or rectangular, since most of the rooms have square or rectangular shape and the room space can easily and naturally be split into square and rectangular communications cells. One should distinguish between the communication cell and the room within which it resides.

The relation between the number of diffusing spots in MSDC and receiver FOV is discussed in Ref. [19]. It was concluded that the optimum number of beams emitted by a transmitter equipped with a computer-generated holographic beam splitter depends on the fabrication accuracy and is usually 100. A natural consideration about the receiver branch FOV is providing each of the receiver branches with signal power, that is, assuring that at least one diffusing spot lies within the FOV of each receiver branch. If ΔS is the diffusing spot grid spacing and R is the radius of the circular area on the ceiling seen by the receiver branch, $R \geq \Delta S / \sqrt{2}$ (FOV$_1$ in Figure 9.4b). The grid spacing is determined by the CGH beam splitter parameters and the distance between the transmitter and the ceiling [20] is given by

$$\Delta S = \frac{\lambda H_t}{\Lambda} \tag{9.1}$$

where λ is the laser wavelength and Λ is the size of the hologram elementary cell. Then the condition for the receiver branch FOV becomes

$$\tan\left(\text{FOV}_1\right) = \frac{R}{H_r} \geq \frac{\Delta S}{\sqrt{2}H_r} = \frac{\lambda H_t}{\sqrt{2}\Lambda H_r} \tag{9.2}$$

where H_r is the distance between the receiver and the ceiling. Notice that the limit of FOV is independent of the cell size, provided that the receiver and the transmitter are positioned at nearly the same height. Also, it is independent of the spot-grid spacing.

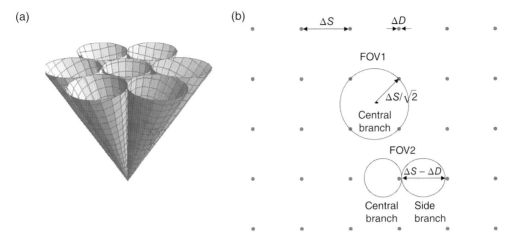

Figure 9.4 (a) Total field-of-view of a seven-branch receiver and (b) areas on the ceiling seen by some of the receiver branches for different field-of-view values.

When this condition holds, a branch can cover more than one diffusing spot. This would cause some deterioration of channel from the ideal state. Other sources of broadening of channel impulse response would be the size of diffusing spot and delayed signal due to multiple reflections. In order to ensure signal receipt from only one diffusing spot, we need to reduce the receiver branch FOV so that $2R < \Delta S - \Delta D$ where ΔD is the diffusing spot diameter.

In MSDC, the receiver usually consists of one central branch aiming at the ceiling and six side branches tilted at an angle twice the FOV of a single branch (Figure 9.4a). In order to ensure that none of the receiver branches will receive light from more than one spot, the condition $2R < \Delta S - \Delta D$ has to be applied to all the branches. If it is fulfilled for the side branches, it will hold for the central branch as well (FOV$_2$ in Figure 9.4b). In this case, the diameter $2R$ of the circular area seen by the central branch has to be replaced in the formula by the major axis of the ellipse seen by a side branch. Thus, the condition for the receiver FOV becomes

$$\tan\left(3\text{FOV}_2\right) - \tan\left(\text{FOV}_2\right) \le \frac{\Delta S - \Delta D}{H_r} = \frac{\lambda H_t}{\Lambda H_r} - \frac{\Delta D}{H_r} \qquad (9.3)$$

If $\Delta D << \Delta S$ and $H_r \approx H_t$, the upper limit for the receiver branch FOV is independent of the cell size and the spot-grid spacing. With $\Delta S = 0.6\,\text{m}$ (10×10 diffusing spots covering $6\,\text{m} \times 6\,\text{m}$ ceiling), $\Delta D = 5\,\text{cm}$, and $H_r = H_t = 2.1\,\text{m}$, the two limits (9.2) and (9.3) for the FOV, corresponding to the two cases of interest when each receiver branch covers at least or at most one diffusing spot, are FOV$_1 \approx 11.5°$ and FOV$_2 \approx 7°$, respectively. In the first case, larger FOV provides more signal power at the expense of bandwidth reduction. The larger bandwidth in the second case comes at the cost of less power efficiency.

We should point out here that although a particular room has been assumed, the conclusions are general and valid as long as the receiver is designed according to (9.2) and (9.3). This arises from the fact that what matters is the match between the receiver FOV and the elevation angles of the emitted beams that create the diffusing spot array. In other words, the values for FOV$_1$ and FOV$_2$ derived above will not change with the size of the communication cell. In a higher communication cell, the same holographic beam-splitter will create the same spot array with the only difference that the spot-grid spacing will be larger. On the other hand, due to the increased cell height, the receiver branch FOV will cover a larger area on the ceiling. As a result, the number of diffusing spots that lie within the receiver branch FOV will not change.

9.4   Receiver Optical Front-End

Narrow FOV of receiver elements reduces the amount of received signal power significantly. Though the optical signal power is concentrated in small-area diffusing spots, eye safety requirements limit the power that can be launched via a single diffusing spot. Hence an optical concentrator has to be used at the receiver. So far as ambient light is concerned, strong sources of ambient light, for example lamps, can be easily rejected using narrow FOV and applying effective combining techniques [21]. However, even a weak diffused background light is much stronger than the optical signal. To improve the optical signal-to-noise ratio, an optical filter that would efficiently reject the optical noise is needed. Conventional receiver optical subsystem consists of a lens concentrator and an optical filter (Figure 9.5a). In this section, the possibility of using a HOE as a receiver branch optical front-end is shown. The main advantages of HOEs over conventional systems are multi-functionality, independence of their physical configurations, insignificant weight, low cost, etc. The performance of different types of holographic mirrors (namely, holographic spherical mirror (HSM) on a spherical substrate, HSM on a flat substrate, and holographic parabolic mirror (HPM) on a flat substrate) are simulated. We have presented the results and the suitability of the mirrors as a receiver optical front-end is discussed.

HOE is independent of its physical configuration, that is, a flat HOE can exhibit concentrating functions such as a conventional curved mirror. It is confirmed by the simulation results, which show insignificant difference in the performance between HSM fabricated on a spherical substrate [22, 23] and HSM fabricated on a flat substrate. Therefore, to prevent redundancy, here we present only the results for spherical and parabolic mirrors, both fabricated on a flat substrate, since the fabrication would greatly be facilitated by using a flat substrate instead of a curved one.

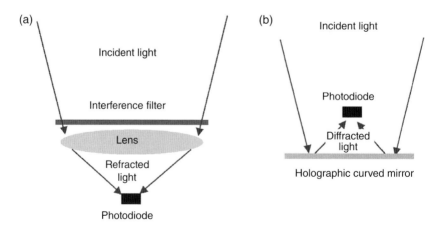

Figure 9.5   (a) Conventional refractive receiver optical front-end and (b) photodiode—holographic mirror system.

9.4.1   Holographic Mirrors

Holographic mirrors are, in fact, reflection holograms [24]. If both recording waves are spherical, the resultant hologram behaves as a spherical mirror. If at recording, one plane and one spherical waves are used, the hologram has the properties of a parabolic mirror. In the simulations, the following parameters of the holograms are assumed—recording wavelength: 850 nm, thickness of the recording medium: 100 µm, refractive index: 1.5, amplitude of the refractive index grating: 0.025, and hologram diameter: 3 cm. When combined with a photo-detector in a single system (Figure 9.5b), the holographic mirror will perform concentration function like a conventional curved mirror. Furthermore, due to its nature, it will diffract light from a very narrow spectral range toward the detector, thus performing filtering function as well. The characteristics of the system, such as FOV, concentration ratio, and spectral bandwidth, depend not only on the hologram characteristics but also on the size and the mutual position of the hologram and the photodetector [22, 23]. The photosensitive area of the photodetector is connected in an inverse manner with the receiver frequency bandwidth through the detector capacitance. A trade-off between the amount of received optical power and the size of the photodetector is always sought. Hence, if two optical front-ends have similar performance, the one serving the smaller photodetector is preferred.

The hologram angular sensitivity imposes difficulties in achieving a FOV greater than a few degrees. This problem can be solved by filling the spacing between the hologram and the detector with a dielectric having the same refractive index as the hologram medium. Furthermore, it reduces the coupling losses at the detector and facilitates system assembly as well.

9.4.2   Signal Effective Area

The received signal radiant flux Φ_S is determined by the hologram diffraction efficiency η and depends on the acceptance angle φ and signal wavelength λ.

$$\Phi_S(\varphi,\lambda) = \oint_S E_S \eta(\varphi,dS,\lambda)\cos\varphi dS = E_S A_{S,\text{eff}}(\varphi,\lambda) \qquad (9.4)$$

where E_S (W cm^{-2}) is the signal irradiance and $A_{S,\text{eff}}$ (cm^2) is the signal effective area of the receiver. The integral is taken over the hologram surface area. We define the cut-off angle of the receiver to be the angle at which the signal effective area becomes zero. The signal effective area dependence on the angle of incidence is shown in Figure 9.6 for the two types of holographic mirrors under investigation for different photodetector size values. Only systems achieving cut-off angles of 12° and 7°, that is, complying with the conditions in (9.2) and (9.3), are compared on the graphs. The optical aberrations in HOEs are stronger than in conventional optics due to their angular and spectral selectivity. The systems achieving larger FOV exhibit significant deterioration in signal reception with an increase in the angle of incidence, especially in the case of HPM. Enlarging the photodetector flattens the angular response of the system and initially increases the amount of received optical power in the case of HSM. However, further increase in the detector size for HSM and any increase in HPM cause a reduction of the received optical power at angles close to normal incidence due to the shadowing caused by the photodetector.

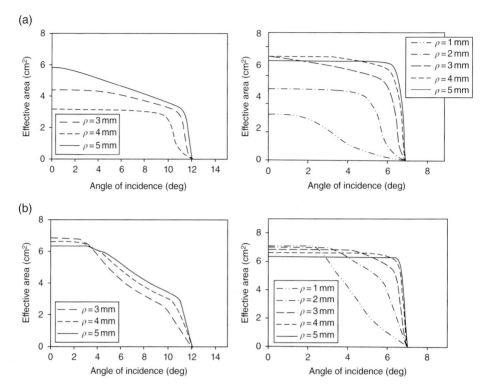

Figure 9.6 Effective area of a photodetector combined with (a) holographic spherical mirror and (b) holographic parabolic mirror for different values of the detector photosensitive area radius ρ. The left graphs represent receiver optical front-end having a cut-off angle of 12°, which corresponds to (9.2), and the right graphs represent the results for an optical front-end with a cut-off angle of 7° that corresponds to (9.3).

9.4.3 Figure-of-Merit

In order to judge the quality of the receiver optical front-end, not only its concentrating capability but also its filtering capability has to be considered. A simple criterion on the optical front-end quality is the improvement in the electrical signal-to-noise ratio compared to a bare photodetector. When shot noise is dominant, the signal-to-noise ratio is determined by the received optical signal and the background radiant flux:

$$\text{SNR}(\varphi) \propto \frac{\Phi_S^2(\varphi)}{\Phi_{\text{bg}}} \tag{9.5}$$

In the case of HOE serving as an optical front-end, the background radiant flux is given by

$$\Phi_{\text{bg}} = \int_{\lambda=0}^{\infty} \int_{\phi=0}^{\pi/2} \oint_S L_{\text{bg}} \eta(\varphi, dS, \lambda) 2\pi \sin\varphi \cos\varphi \, dS \, d\varphi \, d\lambda = L_{\text{bg}} \int_{\lambda=0}^{\infty} (AS)_{\text{bg,eff}} \, d\lambda \tag{9.6}$$

where the spectral background radiance L_{bg} (W cm^{-2} sr^{-1} nm^{-1}) is taken to be a constant over the spectral range within which the diffraction efficiency η is non-zero and $(AS)_{bg,eff}$ (cm^2 sr) is the effective area-solid angle product of the receiver which accounts for the receiver effective area for the ambient light and the effective solid angle within which the ambient light is received.

The electrical signal-to-noise ratio becomes

$$\text{SNR}(\varphi) \propto \frac{\Phi_S^2(\varphi)}{\Phi_{bg}} = \frac{E_S^2}{L_{bg}} \frac{A_{S,eff}^2}{\int_{\lambda=0}^{\infty} (AS)_{bg,eff}\, d\lambda} \tag{9.7}$$

A figure-of-merit can be defined as

$$M(\varphi) = \frac{A_{S,eff}^2}{\int_{\lambda=0}^{\infty} (AS)_{bg,eff}\, d\lambda} \tag{9.8}$$

For a bare photodetector with a photosensitive area of 1 cm^2, the signal-to-noise ratio and the figure of merit are

$$\text{SNR}_{det}(\varphi) \propto \frac{\Phi_S^2(\varphi)}{\Phi_{bg}} = \frac{E_S^2}{L_{bg}} \frac{\cos^2\varphi}{4\pi \sin^2(\text{FOV}/2)\Delta\lambda} \tag{9.9}$$

$$M_{det}(\varphi) = \frac{\cos^2\varphi}{4\pi \sin^2(\text{FOV}/2)\Delta\lambda} \tag{9.10}$$

where $\Delta\lambda$ is the spectral bandwidth of the photodetector. For simplicity, detector responsivity is taken to be constant over 200 nm spectral interval centered on the signal wavelength and to be zero out of this range.

Thus, the improvement in the signal-to-noise ratio with respect to a bare photodiode in decibel is in fact the figure-of-merit gain:

$$G_M(\varphi) = 10\log\left(\frac{A_{S,eff}^2}{\int_{\lambda=0}^{\infty} (AS)_{bg,eff}\, d\lambda} \frac{4\pi \sin^2(\text{FOV}/2)200}{\cos^2\varphi}\right) \tag{9.11}$$

Figure-of-merit gain is presented in Figure 9.7 for different sizes of photodetectors when a holographic spherical or parabolic mirror is utilized. The angular response of the systems with smaller cut-off angle is closer to the ideal rectangular response and allows the utilization of a smaller size photodetector.

When a 12° cut-off angle is needed, the spherical mirror slightly outperforms the parabolic mirror, while the opposite is true in the case of 7° cut-off angle. In the former case, the gain in the electrical signal-to-noise ratio is 13.5 dB at normal incidence and more than 8 dB at the maximum system FOV. In the latter case, it is 18.5 dB at normal incidence and more than 16 dB at the maximum FOV.

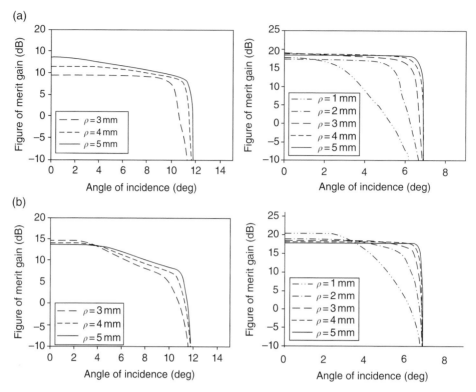

Figure 9.7 Figure-of-merit gain of a photodetector combined with (a) holographic spherical mirror and (b) holographic parabolic mirror for different values of the photodetector radius ρ. The left graphs represent receiver optical front-end having a cut-off angle of 12°, which corresponds to (9.2), and the right graphs represent the results for an optical front-end with a cut-off angle of 7°, which corresponds to (9.3).

9.5 Wave Propagation through Materials and Metamaterials and Relation with Holography

In this section, we digress a bit toward some discussions on metamaterials and their relation with holography as described in this chapter. We discuss what metamaterials are and how they may be useful in holographic applications.

In Ref. [25], we have demonstrated the limited use of cognitive radios and the idea of sharing the radio spectrum through detection and identification of less loaded parts of the frequency spectrum in a congested metropolitan area such as New York, Boston, or Los Angeles. Radio waves, visible light, infrared, ultraviolet, X-rays, and all the other parts of the electromagnetic spectrum are basically electromagnetic radiation, behaving differently in the materials, due to wavelength size. Only about 2 GHz of the lower microwave bands is useful for full wireless mobility due to low absorption and ability to reflect off the objects, thus providing a multiple path channel between a transmitter and a receiver communicating over the channel.

Applications such as Unmanned Aerial Systems require electronic scanning antenna capabilities, in challenging environmental conditions, over very large bandwidths. In addition to that, it is desirable to have as much reduction as possible in size, weight, power, and cost. In mobile transceivers of such systems, antennas are the only components with physical limitations for miniaturization. Steerable antennas using focused beams are able to offer a lot of spatial reuse of the same carrier frequency in order to increase capacity in a mobile cellular network that reuses frequencies. Miniaturization of antennas is limited by physical bounds. Efficient usage of the form factor, effective enlargement of the aperture by coupling to nearby structures and excitation of free space regions, meandering current paths, or engineered (meta)-materials usage—all of these can lead to structures that are as compact as possible.

A material is just a collection of electric and magnetic dipoles. Homogenization allows this collection to be continuous. In other words, it is composed of discrete elements, which are either charged or non-charged particles. If an object (e.g., a single electron, atom, molecule, or particle) is illuminated by an electromagnetic (EM) wave, the interaction between the particle and the EM wave is observed as variations in the traveling EM wave. This interaction is governed by several variables, most prominently, but not limited to, particle size, wavelength, and the relative refractive index of particle to the surrounding medium. This interaction is caused by the excitation of the electric and magnetic dipoles within the object, resulting in an oscillation and radiation of secondary EM waves in what is known as scattering, and/or to transform some part of incident energy into other forms of energy in what is known as absorption [26].

The interaction between particles and EM waves can be fully understood through Maxwell's equations, which provide mathematically convenient forms to evaluate various aspects of the interaction process. From the electromagnetic point-of-view, an atom is just an electric or magnetic, polarizable dipole. Maxwell's equations do not "know" about atoms or molecules; all they "know" is magnetic and electric dipoles. We can use any object to create a dipole response and use that object to form an engineered material.

Material response varies at different frequencies. This is determined by atomic structure and arrangement (10^{-10} m). One can alter a material's electromagnetic properties. A possible method is to introduce periodic features that are electrically small (sub-wavelength sized) over a given frequency range, that appear "atomic" at those frequencies. The paths of light and other electromagnetic waves can be controlled by materials. The lenses in eye glasses or microscopes, for example, are nothing more than pieces of glass or plastic whose surfaces have been shaped in a particular way so as to achieve a desired optical function. Materials are used to form optical devices that operate across the electromagnetic spectrum, from radio waves to visible light. Nature has provided us a rich palette of material properties from which to engineer useful optical devices. Yet, that palette is limited: chemical synthesis, the conventional approach to material development, has so far not enabled us to access the entire range of material properties that should be theoretically possible. But chemistry is not the only process by which we can create materials. As an alternative approach, we can artificially structure a material by assembling a collection of objects together. These objects serve to replace the atoms and molecules of a conventional material, the result being a composite structure that can have electromagnetic properties unlike any naturally occurring or chemically synthesized material. Such composites have been termed "metamaterials," because they have properties that extend beyond materials that occur naturally [27].

New materials, namely metamaterials, are engineered as multi-phase composite materials containing inclusions that often have tailored shapes, sizes, mutual arrangements, and orientations. Such materials exhibit responses unparalleled by many types of wave excitations, including electromagnetic (transverse waves), acoustic (longitudinal waves), and seismic (transverse and longitudinal waves).

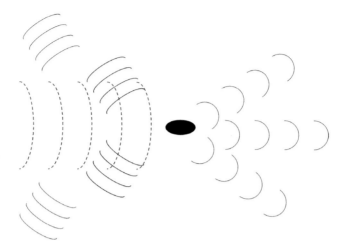

Figure 9.8   Particle–wave interaction; dashed line, incident wave; solid line, scattered wave.

Metamaterials, coined in the 1990s [27], are formed by periodic repetition of some inclusions in a host medium, which may be described as effective media characterized by a set of equivalent constitutive parameters. According to some researchers, any material composed of periodic, macroscopic structures so as to achieve a desired electromagnetic response can be referred to as an engineered-(meta)material. Others prefer to restrict the term metamaterial to materials with electromagnetic properties not found in nature. Almost all agree that the engineered-(meta)materials do not rely on chemical/atomic alterations. Periodic structures and self-similarity are the basis of building volume or 2D holographic components [3, 28]. Similar concepts are used to model phase screens used in modeling the atmospheric turbulence [29]. These concepts have been recognized in some fields for decades. Engineered-(meta)materials are being designed and used in the microwave frequency regime, with possible extension of general concepts into infrared, as Plasmonics [27].

Particles of various densities and properties are present throughout the atmosphere; their distribution varies according to altitude, weather conditions, geographical location, and seasonal changes [30–34]. Thus, it is expected that when an EM wave, referred to as TM or TE modes, traverses between a transmitter and a receiver in the atmosphere, the detected wave will suffer from scattering and absorption (see Figure 9.8).

Spherical particle-wave interaction is commonly known as the Mie theory or the Lorenz Mie theory, in recognition of G. Mie and L. Lorenz who independently developed the principles of electromagnetic plane wave scattering by a dielectric sphere. G. Mie developed the theory in an effort to understand the varied colors in absorption and scattering exhibited by small colloidal particles of gold suspended in water.

Again, the interaction between particles and EM waves can be fully understood through Maxwell's equations. Although it is usually assumed that the particles are spherical, many atmosphere-borne particles are not spherical, but due to the random orientation of the particle in space, particles can be well approximated to be spherical with an average size [30]. Figure 9.9 illustrates the role of sub-wavelength sized particles and the properties of natural materials or engineered metamaterials.

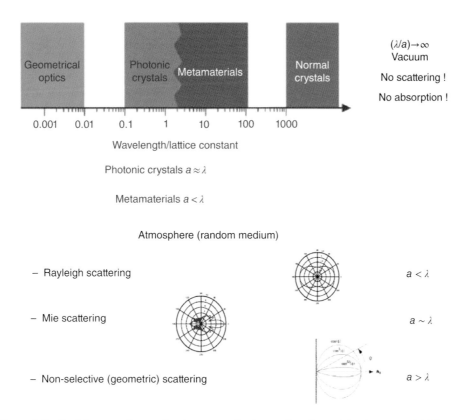

Figure 9.9 Particle–wave interaction in terms of sub-wavelength particle size, a, and EM wave's wavelength, λ.

In vacuum, where the particle size $a=0$, there is no scattering and no absorption; hence, given enough power to make up for a finite transmit/receive aperture antenna size, EM waves can propagate over very large distances carrying extremely broadband signals. If we reverse engineer vacuum, the next best place in terms of propagation distance is within normal crystals (e.g., silicon or glass fiber), where we can have vacuum-like transmissions, with the possibility of launching nonlinear EM modes as Soliton, which is a self-reinforcing (dispersion-free) solitary wave. For the ratio $(\lambda/a) \leq 100$, metamaterials begin to introduce dispersion; thus spatial and frequency selectivity and filtering in space and frequency are observed. Photonic Crystals that exhibit bandgap nature are the extreme examples, as (λ/a) approaches unity. In solid-state physics, a bandgap, also called an energy gap, is an energy range in a solid where no electron states can exist. These materials are known to be least relevant for applications where dispersion is not desirable, for example, fiber optic transmissions.

Natural examples of what photonic crystals are similar to, in behavior, are peacock feathers, rain drops, and butterfly wings that can produce different colors by changing the incidence or view angle with respect to visible light reflection on the object. In imaging, these are referred to as tip and tilt features. For the ranges of $(\lambda/a) \leq 0.01$, the geometric optics range starts where, by adopting ray-tracing, propagation modeling can be performed. A parallel for these behaviors in atmosphere create the Rayleigh scattering (blue sky as seen in the atmosphere) and Mie scattering as light travels through cloud, fog, dust, or smoke particles. Some explosives' smoke, such as fog-oil, exhibits similar scattering of lightwaves.

In 1968, theoretical designs by the Russian scientist Dr. Victor Veselago predicted metamaterials act in the exact opposite manner to natural materials (like negative refractive index) in a pioneering paper [35]. It was not until 1999 that Sir John Pendry [36] explained how to arrive at negative index materials engineered by inclusions of sub-wavelength elements in a substrate to produce such metamaterials. These engineered materials have unique properties not found in nature due to the arrangement and design of the constituents. Metamaterials are causal and dispersive and are called left-handed as they exhibit a negative group phase response resulting in wave vector to have an opposite direction compared to the energy propagation direction. These materials are transparent for propagation, but anisotropic in nature, so the direction dependence they exhibit makes these unsuitable for use in mobile platform antennas, unless an LOS can be guaranteed. However, recent advancements in the introduction of controlled-anisotropy have made it possible to produce frequency selective surfaces that provide tunability and thus scanning and steering features are possible over multiple narrow frequency ranges at a limited slow speed of scanning. The question is, if the anisotropy is controlled, whether these materials can then compete in cost with the cost of conventional materials. Again, LOS between transmitter and receiver is necessary, due to the polarized nature of propagation. These concepts have also resulted in transformational optics (cloaking) and the concept of perfect lens that is capable, in theory, to counter diffraction limit and the losses owing to evanescent fields.

9.6   Conclusions

Today, higher and higher transmission speeds are being pursued for wireless LANs. This chapter has dealt with one of the most promising candidates for high-speed in-house wireless communications, namely, MSDC. As it uses optical medium for data transmission, it possesses inherent potential for achieving very high capacity. We have described a transmitter utilizing a holographic beam splitter for MSDC. The design issues of such a generator of intensity weighted spot arrays have been briefly discussed. The results indicate that two values for the receiver branch field-of-view are worth consideration. We have also discussed novel receiver optical front-end designs and their performance has been analyzed. Taking advantage of the unique properties of the holographic optical elements, the conventional optical front-end, consisting of concentrator and filter, is replaced by a single holographic curved mirror. The use of a HSM improves the signal-to-shot noise ratio by several decibels.

References

[1] G. Yun and M. Kavehrad, "Spot diffusing and fly-eye receivers for indoor infrared wireless communications," in Proceedings of IEEE International Conference on Selected Topics in Wireless Communications, pp. 262–265, Vancouver, BC, June 1992.

[2] E. Simova, M. Tai, and M. Kavehrad, "Indoor wireless infrared link with a holographic multiple-spot diffuser," in Applications of Photonic Technology, G. A. Lampropoulos and R. A. Lessard, eds., Vol. 2. Plenum Press, New York, 1996, pp. 223–228.

[3] M. R. Pakravan, E. Simova, and M. Kavehrad, "Holographic diffusers for indoor infrared communication systems," International Journal of Wireless Information Networks, vol. 4, no. 4, pp. 259–274, 1997.

[4] J. B. Carruthers and J. M. Kahn, "Angle diversity for nondirected wireless infrared communication," in IEEE International Conference on Communications, pp.1665–1670, 1998, vol. 3, Atlanta, GA, June 7–11, 1998, Conference Record.

[5] M. R. Pakravan and M. Kavehrad, "Design considerations for broadband indoor infrared wireless communication systems," *International Journal of Wireless Information Networks*, vol. 2, no. 4, pp. 223–237, 1995.

[6] K.-P. Ho and J. M. Kahn, "Compound parabolic concentrators for narrowband wireless infrared receivers," *Optical Engineering*, vol. 34, no. 5, pp. 1385–1395, 1995.

[7] R. T. Valadas and A. M. de Oliveira Duarte, "Sectored receivers for indoor wireless optical communication systems," in 5th IEEE International Symposium on Personal, Indoor and Mobile Radio Communications, pp. 1090–1095, 1994, vol. 4, Hague, the Netherlands, September 18–22, 1994, Joint PIMRC'94/WCN—Proceedings.

[8] C. R. A. T. Lomba, R. T. Valadas, and A. M. de Oliveira Duarte, "Sectored receivers to combat the multipath dispersion of the indoor optical channel," in Proceedings of the 6th IEEE International Symposium on Personal, Indoor and Mobile Radio Communications, pp. 321–325, 1995, Toronto, Canada, September 27–29, 1995.

[9] A. M. R. Tavares, R. T. Valadas, and A. M. de Oliveira Duarte, "Performance of an optical sectored receiver for indoor wireless communication systems in presence of artificial and natural noise sources," *SPIE*, vol. 2601, pp. 264–273, 1995.

[10] M. Kahn and J. R. Barry, "Wireless infrared communications," *Proceedings of the IEEE*, vol. 85, no. 2, pp. 265–298, 1997.

[11] A. P. Tang, J. M. Kahn, and K.-P. Ho, "Wireless infrared communication links using multi-beam transmitters and imaging receivers," in Proceedings of the IEEE International Conference on Communications, pp. 180–186, 1996, Dallas, TX, June 1996.

[12] J. Kahn, R. You, P. Djahani, A. Weisbin, B. K. Teik, and A. Tang, "Imaging diversity receivers for high-speed infrared wireless communication," *IEEE Communications Magazine*, vol. 361, no. 12, pp. 88–94, 1998.

[13] M. R. Feldman and C. C. Guest, "Iterative encoding of high-efficiency holograms for generation of spot arrays," *Optics Letters*, vol. 14, no. 10, pp. 479–481, 1989.

[14] M. P. Dames, R. J. Dowling, P. McKee, and D. Wood, "Efficient optical elements to generate intensity weighted spot arrays: design and fabrication," *Applied Optics*, vol. 30, no. 19, pp. 2685–2691, 1991.

[15] P. Carnevali, L. Coletti, and S. Patarnello, "Image processing by simulated annealing," *IBM Journal of Research and Development*, vol. 29, no. 6, pp. 569–579, 1985.

[16] P. P. Smyth, D. Wood, S. Ritchie, and S. Cassidy, "Optical wireless: new enabling transmitter technologies," *Proceedings of the IEEE International Conference on Communications*, pp. 562–566, 1993, Geneva, Switzerland, May 1993.

[17] P. L. Eardley, D. R. Wisely, D. Wood, and P. McKee, "Holograms for optical wireless LANs," *IEE Proceedings—Optoelectronics*, vol. 143, no. 6, pp. 365–369, 1996.

[18] F. R. Gfeller and U. H. Bapst, "Wireless in-house data communication via diffuse infrared radiation," *Proceedings of the IEEE*, vol. 67, no.11, pp. 1474–1486, 1979.

[19] S. Jivkova and M. Kavehrad, "Multi-spot diffusing configuration for wireless infrared access: joint optimization of multi-beam transmitter and angle diversity receiver," in Proceedings of the SPIE Conference on Optical Wireless Communications II, SPIE, pp. 72–77, 1999, vol. 3850, Boston, MA, September 1999.

[20] S. Jivkova and M. Kavehrad, "Multi-spot diffusing configuration for wireless infrared access," *IEEE Transactions on Communications*, vol. 48, no. 6, pp. 970–978, 2000.

[21] W. Jeong, M. Kavehrad, and S. Jivkova, "Broadband infrared access with a multi-spot diffusing configuration: performance," *International Journal of Wireless Information Networks*, vol. 8, no. 1, pp. 27–36, 2001.

[22] S. Jivkova and M. Kavehrad, "Holographic spherical mirror as a receiver optical front-end for wireless in-house infrared communications," Applications of Photonic Technology 4, R. Lessard and G. Lampropoulos, Editors, in Proceedings of the SPIE International Conference on Applications of Photonic technology, SPIE pp. 265–273, 2000, vol. 4087, Quebec City, Canada, June, 2000.

[23] S. Jivkova and M. Kavehrad, "Holographic optical receiver front-end for wireless infrared indoor communications," *Applied Optics Journal*, vol. 40, no. 17, pp. 2828–2835, 2001.

[24] R. R. A. Syms, *Practical Volume Holography*. Oxford University Press, New York, 1990.

[25] M. Kavehrad, "Optical wireless applications: a solution to ease the wireless airwaves spectrum crunch," in Invited paper—SPIE Photonics West, San Francisco, CA, February 2–7, 2013.

[26] C. Bohren and D. Huffman, *Absorption and Scattering of Light by Small Particles*. Wiley Interscience, New York, 1983.

[27] D. R. Smith, W. J. Padilla, D. C. Vier, R. Shelby, S. C. Nemat-Nasser, N. Kroll, and S. Schultz, "Left-handed metamaterials," in Costas M. Soukoulis (ed.), *Photonic Crystals and Light Localization*. Kluwer, the Netherlands, 2000.

[28] S. Jivkova and M. Kavehrad, "Indoor wireless infrared local access, multi-spot diffusing with computer generated holographic beam-splitter," in Proceedings of IEEE ICC'99, Vancouver, BC, June 1999.

[29] Z. Hajjarian, J. Fadlullah, and M. Kavehrad, "Use of markov chain in atmospheric channel modeling of free space laser communications," in Proceedings of the IEEE MILCOM, San Diego, CA, November 2008.

[30] F. Smith, *The Infrared and Electro-Optical Systems Handbook, Volume 2: Atmospheric Propagation of Radiation*. SPIE Press, Washington, DC, 1993.

[31] V. De Hulst, *Light Scattering by Small Particles*. John Wiley & Sons, Inc., New York, 1957.

[32] S. Karp, R. M. Gagliardi, S. E. Moran, and L. B. Stotts, eds, "*Optical Channels: Fibers, Clouds, Water and the Atmosphere*. Plenum Press, New York, 1988.

[33] R. M. Harrison and R. E. van Grieken, *Atmospheric Particles, Volume 5*. John Wiley & Sons, Inc., New York, 1998.

[34] J. H. Seinfeld and S. N. Pandis, *Atmospheric Chemistry and Particles*. Wiley Interscience, New York, 1998.

[35] V. Veselago, "The electrodynamics of substances with simultaneously negative values of epsilon and mu," *Soviet Physics Uspekhi*, vol. 10, pp. 509–514, 1968.

[36] J. B. Pendry, A. J. Holden, D. J. Robbins, and W. J. Stewart, "Magnetism from conductors and enhanced nonlinear phenomena," *IEEE Transactions on Microwave Theory and Techniques*, vol. 47, no. 11, pp. 2075–2084, 1999.

10

Indoor Positioning Methods Using VLC LEDs

10.1 Motivation

Indoor positioning is a research area that is gaining lots of attention recently. Its applications cover a wide area where the technology can be incorporated into consumer electronic products. For example, in the case of in-house navigation, indoor positioning technology can be utilized by handheld products to provide location identification, and thus guide users inside large museums and shopping malls, as shown in Figure 10.1a.

Another potential application is location detection of products inside large warehouses where indoor positioning can automate some of the inventory management processes [2, 3]. Indoor positioning techniques, if installed in consumer electronic products, can also be utilized to provide location-based services (LBS) and advertisements to the users. Precise location analytics will also be possible to provide information about consumers' shopping patterns to the retail industry, as depicted in Figure 10.1b. According to a report by the Federal Communications Commission [4], several research results agree that the LBS market size is to be tripled in 2015 compared to that in 2012.

But rapid positioning in indoors is difficult to achieve by global positioning system (GPS) because radio signals from GPS satellites do not penetrate well through walls of large buildings (Figure 10.2). Hence, users having GPS indoors face large positioning errors as well as not being able to connect to GPS satellites at all. To circumvent this situation, two possible alternatives exist, namely radio frequency (RF)-based techniques and visible light communications (VLC)-based techniques.

At the same time, light emitting diode (LED) is being realized as the promising light source in the future. Being a semiconductor light source, LED can be easily modulated for many applications other than lighting, such as indoor communication and smart lighting [5–7]. This feature enables us to explore the possibility to use it for indoor positioning purpose (Figure 10.3).

Short-Range Optical Wireless: Theory and Applications, First Edition. Mohsen Kavehrad, M. I. Sakib Chowdhury and Zhou Zhou.
© 2016 John Wiley & Sons, Ltd. Published 2016 by John Wiley & Sons, Ltd.

(a)

(b)

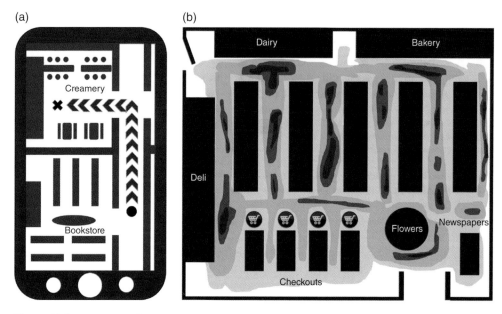

Figure 10.1 Examples of location-based services: (a) indoor navigation for pedestrians and (b) location analytics based on indoor positioning [1].

Figure 10.2 Indoor coverage problem of Global Positioning System (GPS) [1].

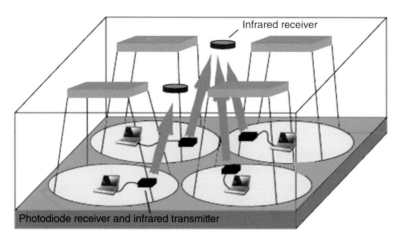

Figure 10.3 Optical wireless communications using visible light LED for downlink and infrared (IR) for uplink [7].

RF-based techniques include, but are not limited to, the following technologies: wireless local area network (WLAN), RF identification (RFID), cellular, ultra-wide band (UWB), and Bluetooth [8–13]. These methods deliver positioning accuracies from tens of centimeters to several meters [14]. But this amount of accuracy is not sufficient for the applications described above. Apart from the relatively poor accuracy of indoor positioning achievable by RF-based techniques, they also add to the electromagnetic (EM) interference. For these reasons, the techniques based on VLC are gaining more attention.

VLC-based techniques employ fluorescent lamps and LEDs. These techniques have the advantage that the sources do not produce EM interference and thus can be used in environments where RF is prohibited. Most of the VLC-based techniques use LEDs as the light source, since they can be modulated more easily compared to fluorescent lamps and, hence, location data can be transmitted in a simpler way. Moreover, LEDs are currently being installed in most buildings, especially larger ones, for example, museums and shopping malls, as the primary lighting source instead of fluorescent lamps since they are advantageous of having much longer life and lower operating cost, though the initial fixed cost is still higher. So, indoor positioning techniques based on VLC and LEDs are excellent options.

This chapter focuses on several fundamental research areas of indoor positioning using VLC technology based on LEDs. We will look into channel multi-access which is one of the most important issues, as there are usually multiple light sources involved competing for the same transmission time. The protocol employed directly determines the system performance in terms of complexity and service outage. We discuss the use of basic framed slotted ALOHA (BFSA) protocol that enables the transmitters (LED bulbs) to work independently without any need of synchronization. The accuracy achieved is also better using VLC technology than RF-based techniques as is shown later in this chapter. We will also investigate the effects of several realistic factors (e.g., installation of LED bulbs and orientation angle of the receiver) to better simulate system performance when deployed into a real indoor environment.

10.2 Positioning Algorithms and Solutions

Investigation in the research field of indoor positioning techniques shows that indoor positioning algorithms proposed to date can all be categorized into three types: Triangulation, Scene analysis, and Proximity [1, 14, 15].

10.2.1 Triangulation

Triangulation is actually the general name of positioning techniques which use the geometric properties of triangles for location estimation. Triangulation has two branches: lateration and angulation. Lateration methods estimate the target location by measuring its distances from multiple reference points. In all the VLC-based positioning systems proposed so far, the reference points are light sources and the optical receiver is colocated with the target. Since it is almost impossible to get the distance directly, we measure received signal strength (RSS) and time of arrival (TOA), or time difference of arrival (TDOA), which are used to mathematically derive the distance. Angulation, on the other hand, measures angles relative to several reference points (angle of arrival, AOA). Then location is estimated by finding the intersection of direction lines, which are radii that range from reference points (light sources) to the target (receiver).

10.2.1.1 Triangulation—Circular Lateration

Circular lateration methods utilize two kinds of measurements: TOA and RSS.

Considering the fact that light speed in air is constant, the distance from the mobile target to reference points will be proportional to the travel time of the light signal. In TOA-based systems, time of arrival measurements with respect to three reference points are needed to locate the target, giving the intersection point of three circles for two-dimensional (2D) positioning, or three spheres for three-dimensional (3D) scenario. An excellent example of TOA-based systems is the widely applied GPS system. In the GPS system, satellites send out navigation messages containing time information in the form of repeating ranging codes and Ephemeris (information of orbits for all satellites). When navigation messages from several satellites are successfully received, circular lateration (or referred to as and used through this chapter, trilateration) is used to calculate the receiver's current location. TOA information can be obtained by using UWB technology and results show that a 2D location can be estimated with a root-mean-square (RMS) error of 14 cm [12].

However, there are two significant problems with TOA-based systems. First, all clocks used by reference points as well as by the target have to be perfectly synchronized. Any inaccuracy in synchronization would be directly transformed into positioning errors. Secondly, there has to be a time stamp included inside the transmitted signal, which potentially requires extra cost in terms of data rate. To satisfy the demands of various indoor applications, positioning accuracy requires ranges from meters to centimeters, which means all clocks in TOA-based systems need to be synchronized at least at the level of several nanoseconds. As a result, the complexity and cost of such systems cannot be properly addressed. Research on VLC positioning utilizing TOA measurements has been very limited. An analysis on the accuracy of TOA-based VLC positioning systems is given in Ref. [16] considering only shot noise.

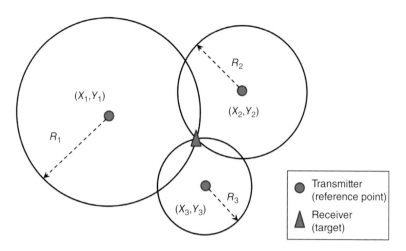

Figure 10.4 Positioning using circular lateration [1].

Simulation results show the Cramer-Rao bound which indicates that, depending on system settings, around 2–5 cm positioning accuracy may be obtained.

RSS-based systems measure the received signal strength and calculate the propagation loss that emitted signal has experienced. After that, range estimation is made by employing a path loss model. Then, circular lateration (trilateration) is performed to estimate the target's position, as shown in Figure 10.4.

Available RSS-based methods using WLAN technology are capable of providing accuracy of 4 m with 90% confidence [11]. VLC-based systems are promised to provide better positioning accuracies due to weaker multipath effects compared to radiowave approaches, leading to better estimation of the RSS. According to simulations done in Ref. [17], the target can be located with an accuracy of around 0.5 mm. In Ref. [18], the authors take the receiver's rotation as well as moving speed into consideration and show by simulations that an overall positioning accuracy of 2.5 cm can be achieved, when the receiver is moving in typical speed.

Now let us focus on mathematical expression of 2D circular lateration. 3D expression is similar. Assume (X_i, Y_i) is the position of the ith reference point (transmitter in most cases) in a 2D space and (x, y) is the position of target (receiver in most cases). If measurement of the distance between the ith transmitter and receiver is given as R_i, then each circle as shown in Figure 10.4 is a set of the receiver's possible locations from a single range measurement, which can be expressed by

$$(X_i - x)^2 + (Y_i - y)^2 = R_i^2 \tag{10.1}$$

where $i = 1, 2, \ldots, n$ and n is the number of reference points. Theoretically, the intersection of all circles described by (10.1) should yield a single point, which is the position of the receiver, given range measurements are not noisy. However, this cannot be the realistic situation. Noises in range measurements lead to a 2D section surrounded by circles given by (10.1). In this case, the approximate solution of the receiver's location is provided by the least squares method, which is discussed in Refs [19, 20].

Note that

$$R_i^2 - R_1^2 = (x - X_i)^2 + (y - Y_i)^2 - (x - X_1)^2 - (y - Y_1)^2 \qquad (10.2)$$

$$= X_i^2 + Y_i^2 - X_1^2 - Y_1^2 - 2x(X_i - X_1) - 2y(Y_i - Y_1) \qquad (10.3)$$

where $i = 1, 2, \ldots, n$. Then the equations describing the system can be transformed into matrix form [20]

$$\mathbf{AX} = \mathbf{B} \qquad (10.4)$$

where

$$\mathbf{X} = [x \quad y]^T \qquad (10.5)$$

$$\mathbf{A} = \begin{bmatrix} X_2 - X_1 & Y_2 - Y_1 \\ \vdots & \vdots \\ X_n - X_1 & Y_n - Y_1 \end{bmatrix} \qquad (10.6)$$

and

$$\mathbf{B} = \frac{1}{2} \begin{bmatrix} \left(R_1^2 - R_2^2\right) + \left(X_2^2 + Y_2^2\right) - \left(X_1^2 + Y_1^2\right) \\ \vdots \\ \left(R_1^2 - R_n^2\right) + \left(X_n^2 + Y_n^2\right) - \left(X_1^2 + Y_1^2\right) \end{bmatrix} \qquad (10.7)$$

The least squares solution of the system is then given by

$$\mathbf{X} = (\mathbf{A}^T \mathbf{A})^{-1} \mathbf{A}^T \mathbf{B} \qquad (10.8)$$

10.2.1.2   Triangulation—Hyperbolic Lateration

Hyperbolic lateration methods usually utilize time-difference-of-arrival (TDOA) measurements. Different from TOA, TDOA-based systems measure the difference in time at which signals from different reference points arrive. These signals are transmitted at exactly the same time; therefore, all the transmitters as reference points have to be synchronized precisely. This can be easily realized because LED bulbs are in close proximity. On the other side, the receiver does not have to be synchronized with transmitters since it is not taking measurements of the absolute TOA. Moreover, no time stamp is required to be labeled in the transmitted signal.

Similar to TOA-based systems, three reference points are needed to perform 2D or 3-D positioning. Since a single TDOA measurement provides a hyperboloid on a 2D plane or a hyperbola in a 3-D space, two TDOA measurements gained with three reference points are needed to locate the target by finding the intersection point. A TDOA-based system using UWB technology experimentally demonstrated a positioning accuracy of about 20 cm [10]. Instead of taking TDOA measurements directly, we may take other measurements first and further compute TDOA information out of them. In Ref. [21], the sinusoidal components of

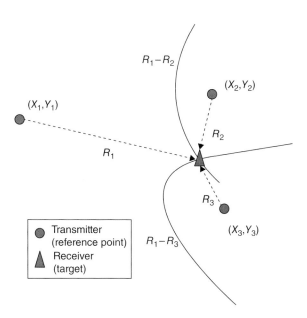

Figure 10.5 Positioning using hyperbolic lateration [1].

signals transmitted from two LEDs are assumed to generate interference pattern at the receiver since they are of the same frequency. Therefore, the peak-to peak value of received sinusoid could be used to derive TDOA measurements. In Ref. [22], phase differences among three signals with different frequencies are used to calculate TDOA information. Computer simulations show the system is capable of delivering an overall accuracy of 1.8 mm.

Following the denotation above, in mathematical expression of 2D hyperbolic lateration, every hyperbola as shown in Figure 10.5 is a set of the receiver's possible locations determined by a single range difference measurement. Every hyperbola can be expressed as

$$D_{ij} = R_i - R_j = \sqrt{(X_i - x)^2 + (Y_i - y)^2} - \sqrt{(X_j - x)^2 + (Y_j - y)^2} \tag{10.9}$$

where D_{ij} denotes the difference between the ranges R_i and R_j, with respect to the ith and jth reference points and $i \neq j$. Note that

$$(R_1 + D_{i1})^2 = R_i^2 \tag{10.10}$$

$$X_i^2 + Y_i^2 - X_1^2 - Y_1^2 - 2x(X_i - X_1) - 2y(Y_i - Y_1) - D_{i1}^2 - 2D_{i1}R_1 = 0 \tag{10.11}$$

where $i = 1, 2, \ldots, n$. Then the equations describing the system can be transformed into matrix form [20] as

$$\mathbf{AX} = \mathbf{B} \tag{10.12}$$

where

$$\mathbf{X} = [x \quad y \quad R_1]^{\mathsf{T}} \tag{10.13}$$

$$\mathbf{A} = \begin{bmatrix} X_2 - X_1 & Y_2 - Y_1 & D_{21} \\ \vdots & \vdots & \vdots \\ X_n - X_1 & Y_n - Y_1 & D_{n1} \end{bmatrix} \tag{10.14}$$

and

$$\mathbf{B} = \frac{1}{2} \begin{bmatrix} \left(X_2^2 + Y_2^2\right) - \left(X_1^2 + Y_1^2\right) - D_{21}^2 \\ \vdots \\ \left(X_n^2 + Y_n^2\right) - \left(X_1^2 + Y_1^2\right) - D_{n1}^2 \end{bmatrix} \tag{10.15}$$

The least squares solution of the system is then given by

$$\mathbf{X} = (\mathbf{A}^{\mathsf{T}}\mathbf{A})^{-1}\mathbf{A}^{\mathsf{T}}\mathbf{B} \tag{10.16}$$

10.2.1.3　Triangulation—Angulation

In AOA-based systems, the receiver measures angles of arriving signals with respect to several reference points. The target is then located by finding the intersection of direction lines (Figure 10.6). Theoretically, only two reference points are needed to perform 2D and three for 3D positioning. The most important feature of AOA-based systems is that no synchronization is needed between reference points and target. In RF domain, AOA technique is mainly applied in UWB systems, and in Ref. [13], the authors propose a joint TOA and AOA estimation algorithm yielding 10–35 cm positioning accuracy, depending on the type of pulse employed. Interestingly, for optical solutions, we can find a historical trace back to older photogrammetry techniques [23] with many similarities. By using imaging receivers, it is relatively easy to

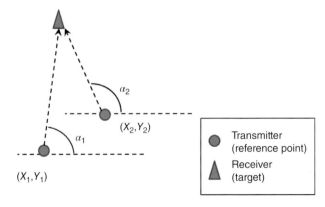

Figure 10.6　Positioning using angulation [1].

detect the AOA of the incoming optical signal. Front-facing cameras on smartphones are inherently imaging receivers. They have brought opportunities to realize AOA-based VLC positioning on consumer electronics. However, current lighting infrastructure has to be adjusted to reach good performance, as most front-facing cameras have a very small field-of-view (FOV). The positioning accuracy will also downgrade when the target moves away from reference points due to the limited resolution of imaging receiver. One available solution to this problem is the use of a fly-eye receiver [24, 25]. As an imaging receiver consists of many independent imaging elements, it provides large FOV while lowering noise on individual channels.

Another problem with AOA methods is the high complexity and cost of imaging receivers, though their size is much smaller than antenna arrays used in radiowave approaches. In Ref. [26], the authors showed that when using an imaging receiver with resolution of 1296×964 pixels, the target can be located with 5 cm accuracy. In Ref. [27], researchers took multipath reflections into consideration and proposed a two-phase positioning algorithm which utilizes both RSS and AOA information. A median accuracy of 13.95 cm can be achieved, according to computer simulations.

To solve the least squares solution of AOA-based system, let α_i denote the AOA measurement with respect to the ith transmitter, which is,

$$\tan \alpha_i = \frac{y - Y_i}{x - X_i} \tag{10.17}$$

where $i = 1, 2, \ldots, n$. We can rewrite (10.17) as

$$(x - X_i)\sin \alpha_i = (y - Y_i)\cos \alpha_i \tag{10.18}$$

Then, we should be able to transform the system equations into matrix form [16],

$$\mathbf{AX} = \mathbf{B} \tag{10.19}$$

where

$$\mathbf{X} = [x \quad y]^{\mathrm{T}} \tag{10.20}$$

$$\mathbf{A} = \begin{bmatrix} -\sin \alpha_1 & \cos \alpha_1 \\ \vdots & \vdots \\ -\sin \alpha_n & \cos \alpha_n \end{bmatrix} \tag{10.21}$$

and

$$\mathbf{B} = \begin{bmatrix} Y_1 \cos \alpha_1 - X_1 \sin \alpha_1 \\ \vdots \\ Y_n \cos \alpha_n - X_n \sin \alpha_n \end{bmatrix} \tag{10.22}$$

Finally the least squares solution of the system is given by,

$$\mathbf{X} = (\mathbf{A}^{\mathrm{T}}\mathbf{A})^{-1}\mathbf{A}^{\mathrm{T}}\mathbf{B} \tag{10.23}$$

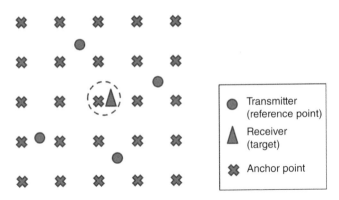

Figure 10.7   Positioning system using scene analysis.

10.2.2   Scene Analysis

Scene analysis refers to a group of positioning algorithms which first collect fingerprints associated with every position in a scene, as shown in Figure 10.7. The target's location is then found by matching real-time measurements to these fingerprints. Factors that can be used as fingerprints include, but are not limited to, all measurements we mentioned earlier, that is, TOA, TDOA, RSS, and AOA. Because of complexity and other concerns, RSS fingerprint is the most used form in both RF and optical domains. Due to the use of fingerprints, the time to complete one single positioning of the target is shorter than that in triangulation methods, thus saving a lot of computation time and power for mobile devices. However, the shortcoming of scene analysis solutions is also apparent. It requires accurate system precalibration for a specific environment and thus cannot be deployed instantly inside a new scenario.

In Ref. [8], the authors built a fingerprint positioning system composed of RFID tags, achieving 1 m accuracy with 90% confidence. Global system for mobile communications (GSM) system can also perform indoor positioning given accurate fingerprint collecting and matching and may provide median accuracy up to 2.84 m, according to experimental results reported in Ref. [9]. A scene analysis method using white LEDs is proposed in Ref. [28]. It utilizes the RSS information from four LED lights as fingerprints for different locations. Results from experiments show that the system delivers a positioning accuracy of 4.38 cm.

10.2.3   Proximity

The working principle of the proximity method is very simple and straightforward. It relies upon a dense grid of reference points, each having a well-known position. When a mobile target receives a signal from a single reference point, it is considered to be colocated with it (Figure 10.8). When signals from multiple sources are received, positions of such sources will be averaged. Therefore, proximity systems using light sources as transmitters theoretically provide accuracy no more than the resolution of the grid itself. Notice that when dense grids

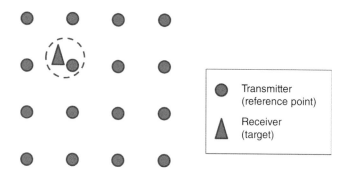

Figure 10.8   Proximity positioning system.

are used, beam profiles of light sources must be optimized in order to minimize inter-source interference, which leads to extra positioning errors. In Refs [2, 3], the authors proposed and experimentally demonstrated a proximity-based indoor positioning system. Apart from positioning provided by the visible light LED, a Zigbee wireless network is used to transmit the location information to the main node, as well as to extend the working range.

10.2.4   Comparison of Positioning Techniques

To compare the systems we have mentioned, we adopt the following performance metrics: accuracy, spatial dimension, and complexity [15].

Accuracy is usually defined as the mean value of positioning error. Here, the positioning error refers to the Euclidean distance between the real position of the target and the estimation. The lower the mean value is, the higher accuracy one system can deliver. As the most important requirement we have on a positioning system, accuracy should be the main factor we consider in judging its performance.

Spatial dimension refers to the dimensions of location information a system can provide. Many solutions proposed so far are capable of providing only 2D positioning, in which case the target is supposed to be located within a horizontal plane. So if elevation of the target changes, accuracy of the system could drop down as a result. Systems featuring 3D positioning, on the other hand, provide better performance in this situation.

The complexity we define here contains two components. One consists of the system requirements on hardware, referring to how many devices are involved and how sophisticated the system configuration is. The hardware complexity will mainly determine what would be the overall cost to deploy an indoor positioning system. The other one is the computing complexity of the positioning algorithm, or in other words, what the delay is before the system is able to update the new location of the target. Since in most of the systems data processing is on the target side, which are the handsets in real scenarios, the algorithm complexity is also important considering the limited battery life of handsets, even though they have very powerful computing capabilities at the present time. Comparison of the algorithms according to these criteria is presented in Table 10.1.

Table 10.1 Comparison of solutions mentioned [1, 15]

Reference	Positioning algorithm	Accuracy	Space dimension	Complexity hardware/ algorithm	Comments
[2]	VLC/proximity	Room level (experiment)	2D	Low/low	4 MHz carrier is employed to realize better optical detection
[8]	RFID/scene analysis	1 m (experiment)	2D	Low/low	Two phases are integrated into the system
[9]	GSM/scene analysis	2.84 m (experiment)	3D	Moderate/ low	Wide signal-strength fingerprints are employed
[10]	UWB/TDOA	20 cm (experiment)	2D	High/ moderate	Use of high speed comparators are compared to 8-bit analog-to-digital converter (ADC)
[11]	WLAN/RSS	4 m (experiment)	2D	Moderate/ moderate	Filtering algorithm is employed
[12]	UWB/TOA	14 cm (simulation)	2D	High/ moderate	Relative location principles are used
[13]	UWB/ TOA+AOA	10 cm (simulation)	2D	High/high	Two pulses yield different accuracies
[17]	VLC/RSS	0.5 mm (simulation)	2D/3-D	Moderate/ moderate	3D positioning is realized
[18]	VLC/RSS	2.5 cm (simulation)	2D	Moderate/ moderate	Rotation and moving speed of the receiver are accounted
[22]	VLC/TDOA	1.8 mm (simulation)	2D	High/ moderate	Frequency-division multiple access (FDMA) protocol employed
[26]	VLC/AOA	4.6 cm (experiment)	3D	High/ moderate	Different colors are used to distinguish LED lights
[27]	VLC/ RSS+AOA	13.95 cm (simulation)	2D/3D	High/high	High Lambertian order sources are assumed ($m = 30$), multipath effects and receiver orientation is taken into consideration
[28]	VLC/scene analysis (RSS based)	4.38 cm (experiment)	2D	Moderate/ low	Offline pre-calibration needed

10.3 An Asynchronous Indoor Positioning System based on VLC LED

As discussed in the last section, in positioning systems based on time measurements (TOA, TDOA), synchronization is a major source of positioning error. In TOA-based systems, it is extremely difficult, if not impossible, to get all transmitters and the receiver to be precisely synchronized. This becomes easier in TDOA-based approaches. The clock used by the receiver does not have to be precisely synchronized with the transmitters and we can make use of other measurements as mentioned in Refs. [21, 22]. Nevertheless, the signals from all transmitters have to be emitted simultaneously. Otherwise, the initial phases of transmitted signals must be measured in a very accurate way. Therefore, synchronization has to be performed among transmitters, which can lead to potential high system deployment cost and limited applicable scenarios. As a result, development of an asynchronous system is of great realistic value.

Furthermore, as implied by the principles of positioning algorithms mentioned earlier, in all systems except proximity based ones, multiple transmitters are required to estimate the receiver position. This means we have to solve the channel multi-access problem to avoid interference. Even for proximity based systems, the problem has to be handled to minimize the possibility of detection failure when multiple transmitted signals are within the receiver FOV if no channel multi-access protocol is employed and transmitters keep on sending out signals all the time.

In GPS systems, channel multi-access is addressed by using code-division-multiple-access (CDMA) technique [29], which allows messages from different satellites to be distinguished from one another. Two types of CDMA encodings are used in the GPS system: the coarse/acquisition (C/A) code accessible by public and the precise (P(Y)) code which is encrypted and only US military forces have access to it. In the TDOA-based system proposed in Ref. [22], frequency-division-multiple-access (FDMA) is selected as a solution. Time-division-multiple-access (TDMA) is another technique used by many systems and requires synchronization among transmitters, resulting in higher deployment cost and time. Here, a new protocol is described to eliminate the need for synchronization. The adoption of BFSA protocol successfully enables the LEDs to work asynchronously, thus no physical connections among them are required, lowering system complexity and cost.

10.3.1 Basic Framed Slotted ALOHA Protocol

10.3.1.1 ALOHA Protocol

ALOHA (Additive Link On-line HAwaii system) protocol [30], which is now referred to as pure ALOHA, was proposed by Norman Abramson and his colleagues at the University of Hawaii to realize connection using low-cost radio equipment instead of conventional wire communications. The protocol was developed for the ALOHA system, also known as ALOHAnet, which demonstrated the first public wireless data network in 1971 [31].

The rules of pure ALOHA are quite simple as follows (Figure 10.9):

1. If a transmitter has a message to send, send the message.
2. The transmitter then waits for an acknowledgement (ACK). If no ACK is received within a time period, it means the message collided with another transmission or message was lost. Wait a random time and then resend the message.

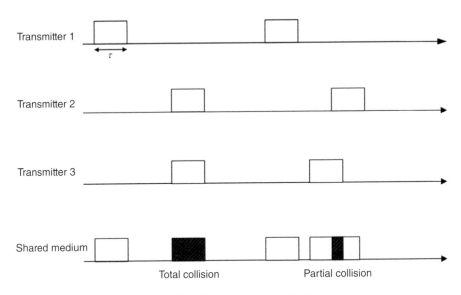

Figure 10.9   Principle of pure ALOHA.

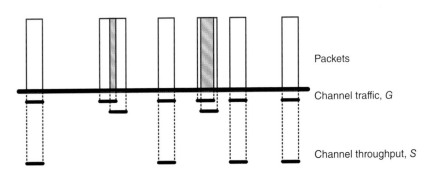

Figure 10.10   Definition of ALOHA traffic G and throughput S.

Pure ALOHA offers poor performance, which is often evaluated by the overall throughput. Throughput is defined as the successful transmissions by all nodes expressed as a fraction of the maximum possible traffic. If we use G to denote the channel traffic, which is the total traffic volume generated by all nodes into the channel versus the maximum possible value, and let S represent the throughput, the relation between them is then given by [32]

$$S = Ge^{-2G} \tag{10.24}$$

This relationship can be shown in a more straightforward way as depicted in Figure 10.10 [32].

From Figure 10.11, we can see that channel throughput reaches its maximum value of 0.184 when $G=0.5$.

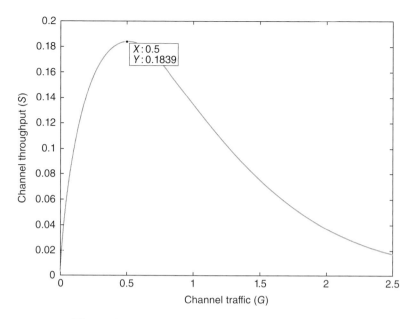

Figure 10.11   Throughput of ALOHA S versus traffic G.

10.3.1.2   S-ALOHA Protocol

To improve channel throughput, slotted ALOHA (S-ALOHA) is proposed [33]. It introduces discrete time slots and users are constrained that they can send at the start of a time slot. Despite the fact that it introduces the need for synchronization into the ALOHA system for better performance compared to the asynchronous version, chances of experiencing collisions are indeed reduced. Notice that if collision does happen in the synchronous S-ALOHA system, one transmission will be totally overlapping with others, as shown in Figure 10.12.

Following the same denotation as in previous discussions, the relation between channel traffic and throughput is given by

$$S = Ge^{-G}$$

(10.25)

From Figure 10.13, we can see that channel throughput reaches its maximum value, which is 0.368, when $G = 1$.

10.3.1.3   BFSA Protocol

BFSA is yet another variant of slotted ALOHA protocol [34]. It defines a frame structure containing a fixed number of timing slots. Each transmitter is allowed to occupy only one slot within a frame length, consequently collisions among transmitters reduce, compared to pure ALOHA. It should also be noted that since the transmitters are not synchronized in the proposed system, the exact start time of transmission in each slot by the transmitters is different, which may lead to overlap among their transmissions even if they select adjacent slots.

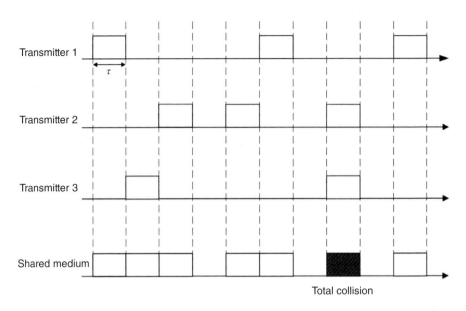

Figure 10.12   Principle of slotted ALOHA.

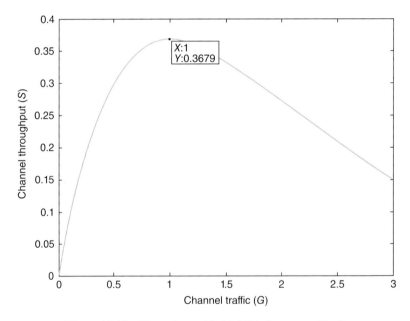

Figure 10.13   Throughput of S-ALOHA S versus traffic G.

In simulations, to obtain the lower bound of system performance, detection failure is supposed as a result of any overlap, even partial. To demonstrate what the real scenario inside the proposed system is, Figure 10.14 shows the working principle of BFSA in the case of having four LED bulbs as transmitters, given no synchronization among them. The figure is drawn according to the clock used by LED Bulb 1.

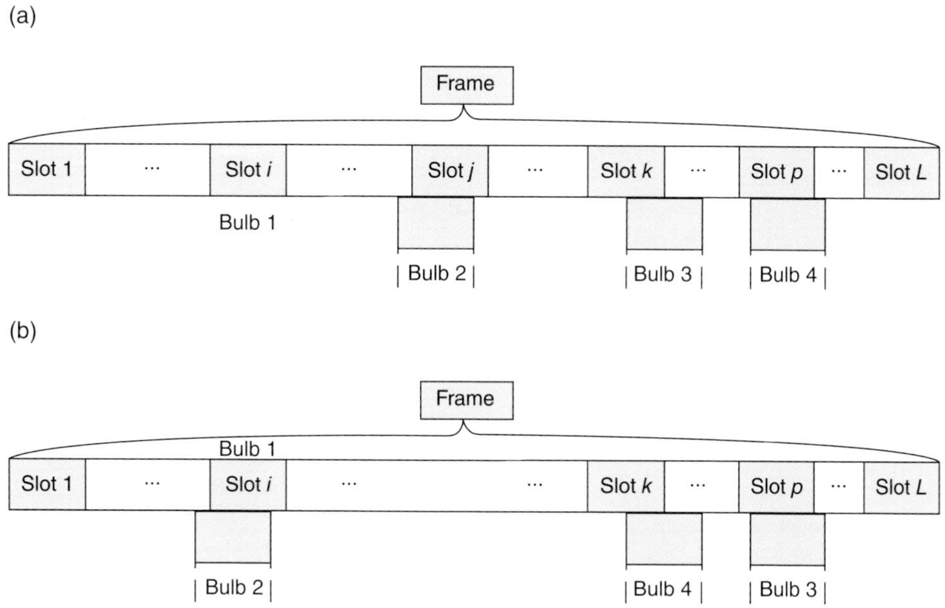

(a)

(b)

Figure 10.14   BFSA protocol: (a) a successful transmission and (b) a transmission failure because of collision.

In the case that every transmitter selects a time slot different from the others as shown in Figure 10.14a, the transmission is defined as a successful one, without any interference, and the receiver can distinguish signals without any problem. If two or more transmitters occupy the same time slot, the receiver will not be able to separate signals due to interference and the transmission is considered to have failed. In the example we show in Figure 10.14b, conflict happens at slot i, since both Bulb 1 and Bulb 2 send out their signals.

The traditional factor used to evaluate the performance of ALOHA-based protocols is throughput. However, it cannot be used in this system for evaluation purposes because the receiver has to receive all signals from different LEDs to estimate its position. In other words, successful reception of only one signal from one transmitter is not enough for the receiver to localize itself. Therefore, the only factor to consider here is the probability of successful transmission from all transmitters to a given receiver.

As mentioned earlier, it is assumed that any overlap among signals will lead to a detection failure. Under this assumption, the probability of successful transmission ($P_{success}$) in BFSA protocol, given total synchronization, is given by

$$P_{success} = \frac{\binom{N}{n}}{N^n} \tag{10.26}$$

where n is the number of transmitters and N is the number of slots in a frame.

When n is bigger than 2, there is no closed mathematical form of $P_{success}$ for asynchronous systems. But difference in performance between synchronous and asynchronous scenarios can

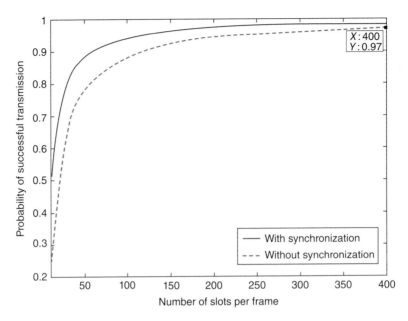

Figure 10.15   Probability of successful transmission versus number of slots per frame, $n=4$.

be obtained by running computer simulations. Figure 10.15 shows $P_{success}$ versus the number of slots per frame, N, for four transmitters. It can be seen that when N is large enough (e.g., $N \geq 200$), the performance difference between synchronous and asynchronous systems is negligible. Also, an overall success rate of 97% for transmissions can be expected if asynchronous setting is employed.

Note that this only gives the performance of the system when signals from four transmitters are within the FOV of the receiver. However, even if we design the FOV in a very accurate way, there may be situations where signals from more than four transmitters are collected by the receiver. We also show here results of simulating the performance of BFSA protocol for eight transmitters, that is, $n=8$ in Figure 10.16. As we see, an average success rate of 86.8% can be reached in the presence of eight transmitters, which is still acceptable.

A drawback of BFSA protocol is that it requires bandwidth for individual transmitters to be N times of what was originally needed. However, the data rate this system is working at is not very high; so, a relatively large N can be chosen to ensure successful transmissions for most of the time. Suppose 128 bits are used to represent the hardware ID for individual LED bulbs, plus an 8-bit beginning flag, an 8-bit ending flag, and a segment of 16-bit CRC to form a packet that is to be transmitted within a frame. The system will perform positioning 10 times per second and use a frame structure containing 400 slots. The required data rate is then given by

$$400 \times (128 + 8 + 8 + 16) \text{ bit} / 0.1 \text{ s} = 640 \text{ kbps} \qquad (10.27)$$

This speed is definitely deliverable even with available off-the-shelf LEDs. On the other hand, with 128 bits, the system will be able to label $2^{128} = 3.4 \times 10^{38}$ LEDs, which is far more

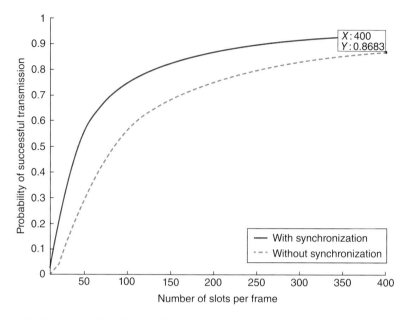

Figure 10.16   Probability of successful transmission versus number of slots per frame, $n=8$.

than actually needed. Therefore, the proposed indoor positioning system is a future-proof solution that works universally at every place where supporting LEDs are installed.

10.3.2   System Design and DC Channel Gain

As shown in Figure 10.17, the system contains a set of LED bulbs on the ceiling and a mobile terminal, which can be either consumer electronics as shown or dedicated devices. LED bulbs function as transmitters, sending different hardware IDs related to their physical locations. An optical receiver inside the mobile terminal extracts the location information to perform positioning.

The optical channels concerned here are all line-of-sight (LOS) links. The most important quantity to characterize such links is the DC (direct current) gain. In LOS links, the channel gain can be estimated fairly accurately by considering only the LOS propagation path. The channel DC gain is expressed as

$$H(0) = \begin{cases} \dfrac{m+1}{2\pi d^2} A\cos^m(\phi)T_S(\psi)g(\psi)\cos(\psi), & 0 \le \psi \le \Psi_C \\ 0, & \psi > \Psi_C \end{cases} \tag{10.28}$$

where A is the physical area of the detector, ψ is the angle of incidence with respect to the receiver axis, $T_S(\psi)$ is the gain of optical filter, $g(\psi)$ is the concentrator gain, Ψ_C is the concentrator FOV semi-angle, ϕ is the angle of irradiance with respect to the transmitter perpendicular axis, and d is the distance between transmitter and receiver, as shown in Figure 10.18.

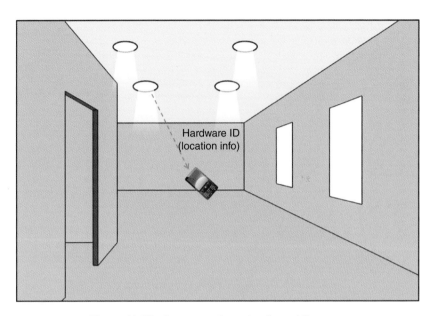

Figure 10.17   System configuration for mobile users.

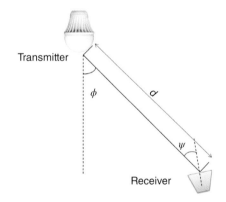

Figure 10.18   Channel model in the simulation.

The Lambertian order m is given by $m = -\ln 2 / \ln(\cos \Phi_{1/2})$, where $\Phi_{1/2}$ is the half power angle of the LED bulb. The received optical power is given by $P_r = H(0)P_t$, where P_t denotes the transmitted optical power from the LED bulb.

10.3.3   *Positioning Algorithm*

10.3.3.1   **Optical Concentrators**

To maximize optical power collected by the receiver, P_r, optical concentrators can be used. Generally, optical concentrators can be divided into two kinds: imaging and non-imaging. Imaging optical concentrators are generally used in directed LOS links. The focal points

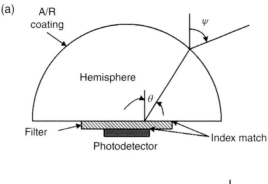

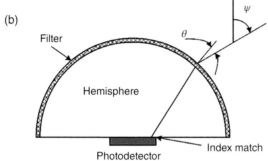

Figure 10.19 Hemispherical optical concentrators: (a) with planar optical filter and (b) with hemispherical optical filter.

generated by imaging concentrators do not change position when light arrives with different angles of incidence. However, the FOV of imaging receivers is very limited, preventing them from being used in non-directed scenarios.

There are three major non-imaging optical concentrators in use for optical communications at this time: hemisphere lens, compound parabolic concentrator (CPC), and dielectric total internal reflecting concentrator (DTIRC). Hemispherical concentrators are easy to fabricate and come with very wide FOV. They also provide concentrator gain equal to n_C^2, where n_C is the refractive index of the concentrator's material.

However, a huge drawback of hemispherical concentrators is that they cannot be coupled with planar thin-film bandpass filters as shown in Figure 10.19a [35]. The reason behind this is that θ, the angle at which light strikes the filter, changes when angle of incidence ψ shifts. This leads to a shift in the filter passband, decreasing the filter transmission $T_s(\psi)$ for certain ψ values. Therefore, it is preferred that bandpass filter is deposited or bonded in a hemispherical shape onto the concentrator, as shown in Figure 10.19b.

The basic concept of CPC was first explained by Welford and Winston in Ref. [36]. CPC-based receivers can be designed to have any FOV ranging from 0 to 90°, though usually they are designed to have limited FOV. When FOV is smaller than 90°, the concentrator gain is expected to exceed n_C^2, which can be provided by hemispherical concentrators. Also, CPCs can be coupled very well with planar optical bandpass filters, which can be fabricated very easily with current techniques, as shown in Figure 10.20a. To achieve large FOV close to 90°, an inverted CPC can be placed in front of the optical filter, as shown in Figure 10.20b and c [37].

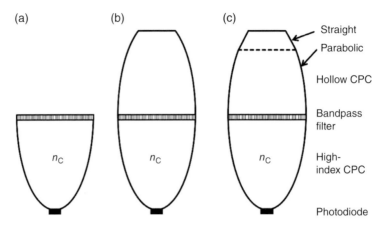

Figure 10.20 CPCs combined with bandpass filters with different FOVs: (a) single dielectric CPC yielding small FOV; (b) CPC coupled with an inverted hollow CPC, achieving FOV of 90°; and (c) CPC coupled with an inverted hollow CPC with straight part, achieving FOV less than 90°. Adapted from Ref. [37].

DTIRC was introduced by Ning *et al.* in 1987 [38]. DTIRCs are promised to provide concentration gain close to the theoretical maximum value. Compared to CPCs, DTIRC-based receivers provide higher concentration gain and smaller size. However, most DTIRCs have limited FOVs as pointed in Ref. [39]. Thus, we assume the use of CPC receiver in the simulations.

10.3.3.2 Triangulation Using RSS Information

In the system, the RSS information of received signal will be used to estimate the receiver's distances from transmitters on the ceiling, after which the receiver will be located by triangulation.

Each of the LED bulbs will simply transmit its own code, modulated in on-off-keying (OOK) format. An LED bulb will use only one slot within the duration of one frame length, while delivering output at a constant power level for illumination purpose. Therefore, the receiver will be receiving only one modulated signal at one instance for most of the time. The OOK modulation used in this system has a modulation depth of 12.5% to minimize the flickering problem [40]. Since the optical emitted power is linearly proportional to the amplitude of the electrical signal, the difference in transmitted power between logical 0s and 1s at the transmitter side is given as

$$P_{\text{diff}} = \eta_{\text{OOK}} P_{\text{const}} \tag{10.29}$$

where η_{OOK} is the modulation depth of the OOK modulation and equal to 0.125 in the simulation. P_{const} is the optical power emitted from LED bulb without modulation. Therefore, the difference at the receiver side $P_{\text{diff\_r}}$ will be given by

$$P_{\text{diff\_r}} = H(0)P_{\text{diff}} = \frac{m+1}{2\pi d^2} A \cos^m(\phi) T_S(\psi) g(\psi) \cos(\psi) P_{\text{diff}} \tag{10.30}$$

Given the receiver's FOV is large enough so that $0 \leq \psi \leq \Psi_C$ always holds, the distance between the transmitter and the receiver, d_{est}, can be estimated by measuring P_{diff_r} at the receiver [41, 42],

$$d_{est} = \sqrt{\frac{(m+1)A\cos^m(\phi)T_S(\psi)g(\psi)\cos(\psi)P_{diff}}{2\pi P_{diff\_r}}} \qquad (10.31)$$

The optical concentrator gain $g(\psi)$ for a CPC is given as [35]

$$g(\psi) = \begin{cases} \dfrac{n_C^2}{\sin^2(\Psi_C)}, & 0 \leq \psi \leq \Psi_C \\ 0, & \psi > \Psi_C \end{cases} \qquad (10.32)$$

where n_C denotes the refractive index of the concentrator.

Assuming both the receiver axis and the transmitter axis to be perpendicular to the ceiling, the following equations hold:

$$d_{est} = \sqrt{d_{est-xy}^2 + H^2} \qquad (10.33)$$

$$\cos(\phi) = \cos(\psi) = \frac{H}{d_{est}} \qquad (10.34)$$

where d_{est-xy} is the estimated horizontal distance between the transmitter and the receiver and H is the vertical distance between the ceiling and the receiver. All the LEDs can be characterized as first-order Lambertian light sources; so, $m=1$. And, to simplify the calculation, the transmission of optical filter and the gain of optical concentrator are combined into one gain:

$$T_S(\psi)g(\psi) = G \qquad (10.35)$$

where G is a constant related to characteristics of filter and concentrator. Consequently,

$$d_{est-xy} = \sqrt{\sqrt{\left(\frac{A \cdot G \cdot P_{diff} \cdot H^2}{\pi \cdot P_{diff\_r}}\right)} - H^2} \qquad (10.36)$$

So long as the positions of LEDs related to different codes are known to the receiver, by collecting signals coming from at least three LEDs, the receiver will be able to use triangulation to determine its current position on a 2D plane, as shown in Figure 10.21.

10.3.3.3 Linear Least Squares Estimation

To estimate the unknown position of the receiver by knowing the distances from several reference points (transmitters' horizontal coordinates), least squares estimation can be employed. The notion "least squares" means that the sum of the "squares" of errors in the results of every

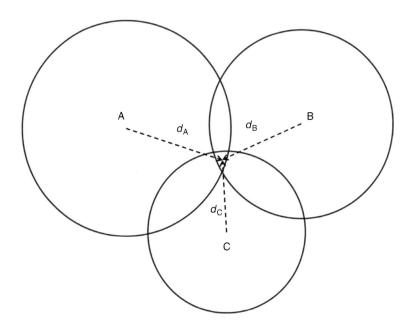

Figure 10.21   Circular lateration.

known equation is minimized by the provided solution. It applies to estimation and reconstruction problems where the relation between the measurements and the unknowns is

$$\mathbf{AX} = \mathbf{B} \tag{10.37}$$

where $\mathbf{X}$ is what we want to estimate or reconstruct; $\mathbf{B}$ is our observation, for example, data from sensors; and $\mathbf{A}$ represents the characteristics of our observations, for example, ith row of $\mathbf{A}$ characterizes ith sensor. Here, $\mathbf{A} \in \mathbb{R}^{p \times q}$ and $p > q$. This is often referred to as overdetermined set of equations, meaning there are more equations than unknowns. Least squares estimation is then defined by choosing $\mathbf{X}$ that minimizes

$$r = \|\mathbf{AX} - \mathbf{B}\| \tag{10.38}$$

where r is called the residual or error.

Least squares solutions can be divided into two kinds: linear least squares (LLS) estimators and nonlinear least squares estimators. Although nonlinear least squares estimators can achieve optimal positioning accuracy, global convergence cannot be guaranteed due to their multi-modal optimization cost functions [43]. On the other hand, LLS is capable of providing global solution because it converts nonlinear equations into linear ones, which usually have global convergence. In most cases, an extra range measurement is required and that is the reason why at least three range measurements are needed for circular lateration. In the simulation, LLS estimation is used since it will provide the most reliable estimation when there are a small number of reference points.

After the estimated horizontal distance between the receiver and each of the three transmitters (denoted as A, B, and C) is obtained, a set of three quadratic equations as follows is formed:

$$(x - x_A)^2 + (y - y_A)^2 = d_A^2 \tag{10.39}$$

$$(x - x_B)^2 + (y - y_B)^2 = d_B^2 \tag{10.40}$$

$$(x - x_C)^2 + (y - y_C)^2 = d_C^2 \tag{10.41}$$

where $[x_A, x_B, x_C]$ and $[y_A, y_B, y_C]$ are the coordinates of LED bulbs in x and y axes, $[d_A, d_B, d_C]$ are the horizontal distances from the receiver to LED bulbs, and (x, y) is the receiver's position to be estimated.

LLS solution is calculated as follows. First, we eliminate the nonlinear part by subtracting (10.41) from the other two equations, which gives us,

$$\begin{cases} 2x(x_A - x_C) + x_C^2 - x_A^2 + 2y(y_A - y_C) + y_C^2 - y_A^2 = d_C^2 - d_A^2 \\ 2x(x_B - x_C) + x_C^2 - x_B^2 + 2y(y_B - y_C) + y_C^2 - y_B^2 = d_C^2 - d_B^2 \end{cases} \tag{10.42}$$

Then, to get one estimation for the receiver's position (x, y), the following matrix form can be used for calculation:

$$\mathbf{X} = [x \quad y]^T \tag{10.43}$$

$$\mathbf{A} = \begin{bmatrix} x_B - x_A & y_B - y_A \\ x_C - x_A & y_C - y_A \end{bmatrix} \tag{10.44}$$

and

$$\mathbf{B} = \frac{1}{2} \begin{bmatrix} \left(d_A^2 - d_B^2\right) + \left(x_B^2 + y_B^2\right) - \left(x_A^2 + y_A^2\right) \\ \left(d_A^2 - d_C^2\right) + \left(x_C^2 + y_C^2\right) - \left(x_A^2 + y_A^2\right) \end{bmatrix} \tag{10.45}$$

The least squares solution of the system is then given by

$$\mathbf{X} = (\mathbf{A}^T \mathbf{A})^{-1} \mathbf{A}^T \mathbf{B} \tag{10.46}$$

Due to linearization involved as introduced below and observation noise, one cannot expect an exact solution satisfying (10.37). However, least squares still provide the best linear unbiased estimator (BLUE) because for a noisy linear observation

$$\mathbf{B} = \mathbf{AX} + \mathbf{V} \tag{10.47}$$

Consider a linear estimator

$$\hat{\mathbf{X}} = \mathbf{BY} \tag{10.48}$$

The estimator is called unbiased if $\hat{\mathbf{X}} = \mathbf{X}$ whenever $\mathbf{V} = 0$; this is equivalent to $\mathbf{BA} = \mathbf{I}$, that is, $\mathbf{B}$ is left inverse of $\mathbf{A}$.

Estimation error of such an unbiased linear estimator is

$$\mathbf{X} - \hat{\mathbf{X}} = \mathbf{X} - \mathbf{B}(\mathbf{AX} + \mathbf{V}) = -\mathbf{BV} \tag{10.49}$$

It is a fact that $\mathbf{A}^{\dagger} = (\mathbf{A}^{\mathrm{T}}\mathbf{A})^{-1}\mathbf{A}^{\mathrm{T}}$ is the smallest left inverse of $\mathbf{A}$, that is, for any $\mathbf{B}$ satisfying $\mathbf{BA} = \mathbf{I}$, we will have

$$\sum_{i,j} B_{ij}^2 = \sum_{i,j} A_{ij}^{\dagger 2} \tag{10.50}$$

Therefore we have proved that the solution given by (10.46) is the BLUE.

If more than three reference points are involved, (10.44) and (10.45) will have more rows than what is shown above. In the simulation, all the equation groups formed by any three quadratic equations are first solved, after which the estimations are processed to obtain the final estimated position of the receiver.

10.3.4 Signal-to-Noise Ratio Analysis

The system configuration is depicted in Figure 10.22. As shown, there are four LED bulbs located on the ceiling. Since the LEDs within one bulb are colocated and modulated by the same circuit, they will perform as a single optical transmitter.

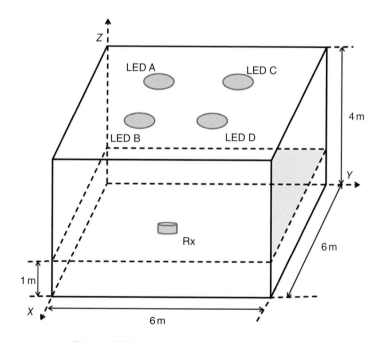

Figure 10.22 System model used in simulation.

Table 10.2 Parameters in simulation

Parameter	Value
Room dimension ($L \times W \times H$)	6 m × 6 m × 4 m
Power of LED bulb (P_{const})	16 W each
Positions of LED bulbs (x, y, z) (m)	A (2, 2, 4)
	B (4, 2, 4)
	C (2, 4, 4)
	D (4, 4, 4)
Codes used by LED bulbs	A (0 1 1 1)
	B (1 0 1 1)
	C (1 1 0 1)
	D (1 1 1 0) (the rest 124 bits are same)
Modulation depth (η_{OOK})	12.5%
Modulation bandwidth (B)	640 kHz
Receiver height	1 m

In the proposed system, each of these four LED bulbs will transmit a unique code assigned to them (first four bits are 0111, 1011, 1101, 1110 respectively, the remaining 124 bits are similar), modulated in OOK format. The receiver is located within the horizontal plane of 1 m height, facing up perpendicularly to the ceiling. Parameters used in the simulation are shown in Table 10.2.

Before running the simulation, the signal-to-noise-ratio (SNR) is calculated to examine the effects of it on the system. The signal component is given by

$$S = \gamma^2 P_{diff\_r}^2 \tag{10.51}$$

where γ is the detector responsivity and P_{diff_r} is given by (10.30).

Following Komine and Nakagawa's research [44], the noise is Gaussian having a total variance N, which is the sum of contribution from shot noise and thermal noise, given by

$$N = \sigma_{shot}^2 + \sigma_{thermal}^2 \tag{10.52}$$

The shot noise variance is given by

$$\sigma_{shot}^2 = 2q\gamma P_{rec} B + 2qI_{bg}I_2 B \tag{10.53}$$

where q is the electronic charge, B is equivalent noise bandwidth, which is equal to the modulation bandwidth here, I_{bg} is background current, whose traditional value is 5100 μA given direct sunlight exposure and 740 μA assuming indirect sunlight exposure [45], and the noise bandwidth factor, $I_2 = 0.562$ [44]. A p-i-n/FET (field-effect transistor) transimpedance receiver is used and the noise contributions from gate leakage current and $1/f$ noise are negligible [46].

P_{rec} is given by

$$P_{rec} = \sum_{i=1}^{4} H_i(0)P_i \tag{10.54}$$

Table 10.3 Receiver parameters

Parameter	Value
Field of view (Ψ_C) (half angle)	70°
Physical area of photodetector (A)	1.0 cm²
Gain of optical filter ($T_S(\psi)$)	1.0
Refractive index of optical concentrator (n_C)	1.5
O/E conversion efficiency (γ)	0.54 A/W

where $H_i(0)$ and P_i are the channel DC gain and instantaneous emitted power for the ith LED bulb, respectively.

On the other hand, the thermal noise variance is given by

$$\sigma_{thermal}^2 = \frac{8\pi k T_K}{G_o} \eta A I_2 B^2 + \frac{16\pi^2 k T_K \Gamma}{g_m} \eta^2 A^2 I_3 B^3 \qquad (10.55)$$

where the two terms represent feedback-resistor noise and FET channel noise. Here, k is the Boltzmann's constant, T_K is the absolute temperature, G_o is the open-loop voltage gain, η is the fixed capacitance of photo detector per unit area, Γ is the FET channel noise factor, g_m is the FET transconductance, and $I_3 = 0.0868$.

In the simulation, the following values are used [47]: $T_K = 295$ K, $G_o = 10$, $g_m = 30$ mS, $\Gamma = 1.5$, $\eta = 112$ pF cm⁻². Parameters of the receiver are listed in Table 10.3.

To evaluate the impact of SNR on the system, a distribution map of SNR with respect to LED bulb D, which is located at (4 m, 4 m, 4 m), is shown in Figure 10.23, assuming direct sunlight exposure, and in Figure 10.24, assuming indirect sunlight exposure.

As can be observed from the figures, the SNR is about 10 dB lower in direct sunlight-exposed room environment for the same location, compared to indirect sunlight exposure. This is mainly because of the increased contribution from shot noise to the total noise level. Besides, it is shown in both figures that the signal transmitted from LED bulb D experiences a very low SNR at the room corner (0 m, 0 m, 0 m). This will be translated into relatively large error in estimating the distance between the LED bulb D and the receiver when it is located at the corner, in other words, the system performance will be downgraded when the user is far away from the LED bulbs.

10.3.5 Results and Discussions

The system performance is evaluated when the receiver is at a fixed height of 1 m. 0.05 m is set as the resolution for the receiver positions, meaning that the Euclidean distance between adjacent positions of the receiver is 0.05 m. Figure 10.25 shows the positioning error distribution in the presence of direct sunlight exposure and Figure 10.26 depicts the result when indirect sunlight exposure is assumed.

As expected, the positioning errors are relatively small for the majority of the room area, but become significantly large when the receiver approaches corners. Moreover, the positioning errors obtained with indirect sunlight exposure are much smaller compared to direct sunlight

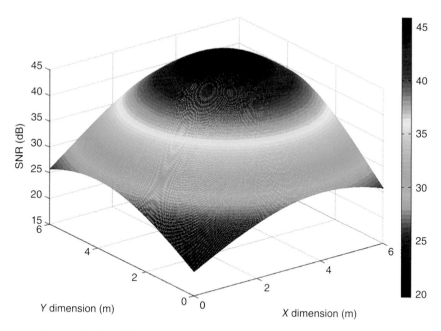

Figure 10.23 SNR distribution for LED bulb D (direct sunlight exposure).

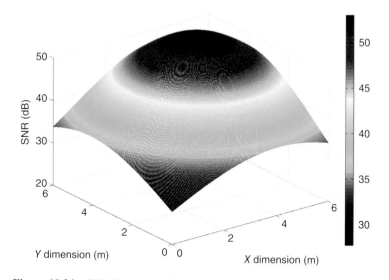

Figure 10.24 SNR distribution for LED bulb D (indirect sunlight exposure).

exposure situation, which is a noisier environment. To directly compare the difference between these two scenarios, the histograms of positioning errors are shown, respectively, in Figures 10.27 and 10.28.

To assess the performance of a positioning system more practically, precision is widely employed. Precision indicates how the system consistently delivers a certain level of service

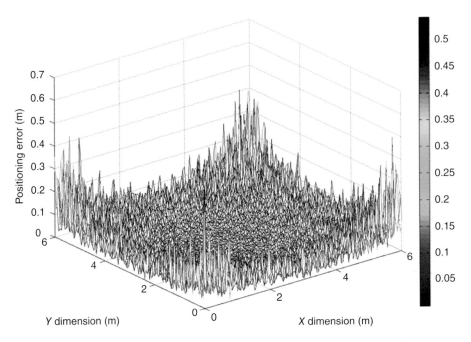

Figure 10.25 Positioning error (in units of meter) distribution under direct sunlight exposure.

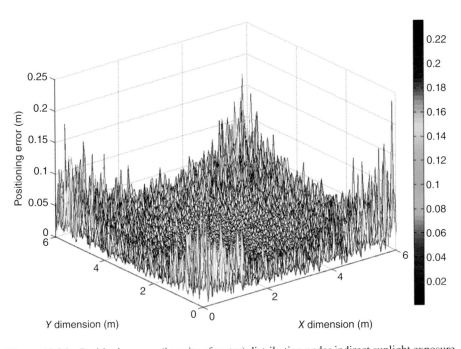

Figure 10.26 Positioning error (in units of meter) distribution under indirect sunlight exposure.

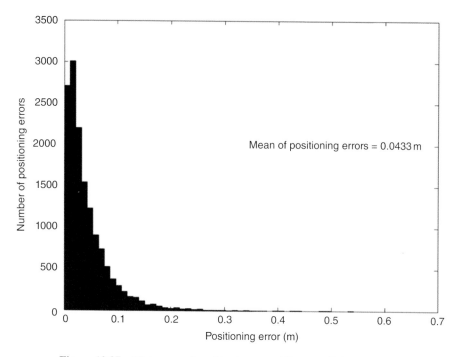

Figure 10.27 Histogram of positioning error (direct sunlight exposure).

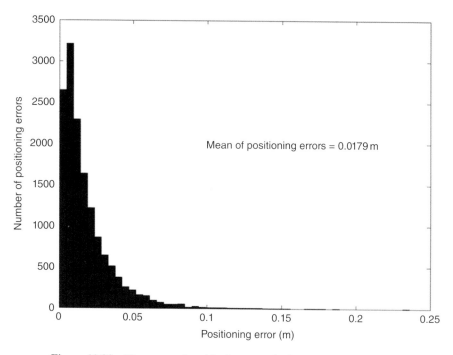

Figure 10.28 Histogram of positioning error (indirect sunlight exposure).

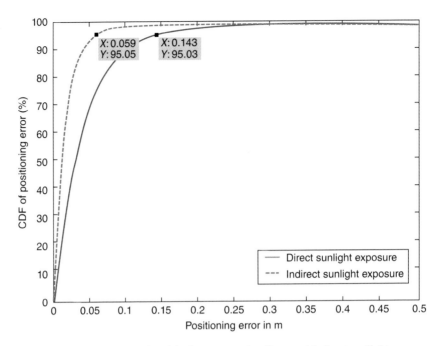

Figure 10.29   CDF curves of positioning error under direct and indirect sunlight exposure.

within a long time-scale (i.e., over many trials). Generally, the cumulative distribution function (CDF) of positioning error is used to evaluate the precision. By taking the performance of BFSA protocol into consideration, using a frame structure containing 400 slots per frame for simulation, the precision curve (CDF curve of positioning error) of this system is obtained and shown in Figure 10.29.

As indicated by the curves, if 95% is assumed as an acceptable service coverage rate, the proposed system will be able to deliver an accuracy of 14.3 cm (5.9 cm), within indoor environments with direct (indirect) sunlight exposure.

10.3.6   Extended Simulation and Results

An extended simulation was completed to further evaluate the system performance under realistic conditions, taking wrong positioning of LED bulbs as well as orientation angle of the receiver into consideration. The parameters used in this extended simulation are shown in Table 10.4.

Figures 10.30 and 10.31 show the positioning error distribution assuming wrong positions of LED bulbs, in the presence of direct and indirect sunlight exposure, respectively. As we can see in these figures, the accuracy of positioning is not severely downgraded due to wrong positioning of LED bulbs and orientation angle of the receiver. This conclusion is confirmed by Figures 10.32 and 10.33, which show the histogram of positioning error under both direct and indirect sunlight exposure.

As we can see, the mean of positioning errors, in both situations, increases compared to the values shown in Figures 10.27 and 10.28, but not dramatically. Therefore, the performance of

Table 10.4 Parameters in extended simulation

Parameter	Value
Room dimension $(L \times W \times H)$	$6\,m \times 6\,m \times 4\,m$
Power of LED bulb (P_{const})	16 W each
Positions of LED bulbs (x, y, z) (m)	A (2, 2, 4.005)
	B (3.98, 2, 4)
	C (2.01, 4.005, 4)
	D (4, 4.01, 3.99)
Codes used by LED bulbs	A (0 1 1 1)
	B (1 0 1 1)
	C (1 1 0 1)
	D (1 1 1 0) (the remaining 124 bits are the same)
Modulation depth (η_{OOK})	12.5%
Modulation bandwidth (B)	640 kHz
Maximum orientation angle of the receiver	5°
Receiver height	1 m

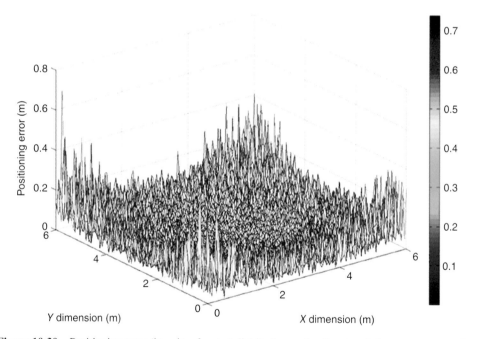

Figure 10.30 Positioning error (in units of meter) distribution under direct sunlight exposure assuming wrong positions of LED bulbs and orientation angle of the receiver.

this system will remain within the same level, even if the LED bulbs involved are installed at slightly incorrect positions.

The precision curves (CDF curve of positioning error) of this system assuming wrong positions of LED bulbs are shown in Figure 10.34. As indicated by the curves, if 95% is assumed

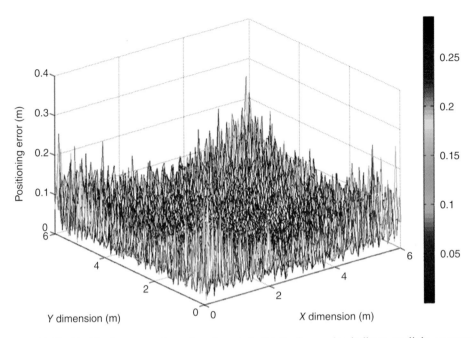

Figure 10.31 Positioning error (in units of meter) distribution under indirect sunlight exposure assuming wrong positions of LED bulbs and orientation angle of the receiver.

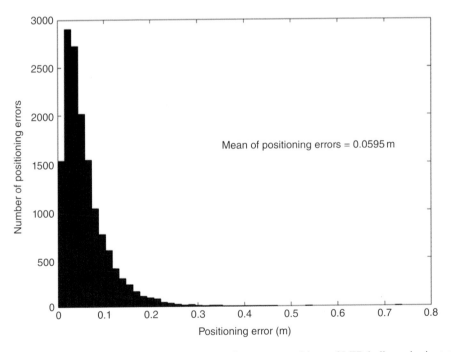

Figure 10.32 Histogram of positioning error assuming wrong positions of LED bulbs and orientation angle of the receiver (direct sunlight exposure).

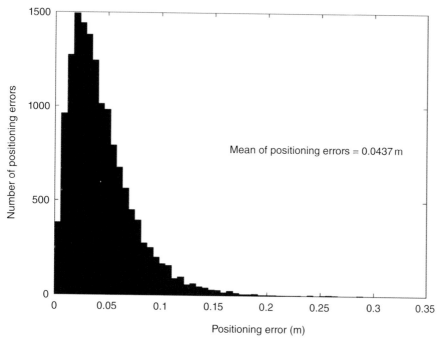

Figure 10.33 Histogram of positioning error assuming wrong positions of LED bulbs and orientation angle of the receiver (indirect sunlight exposure).

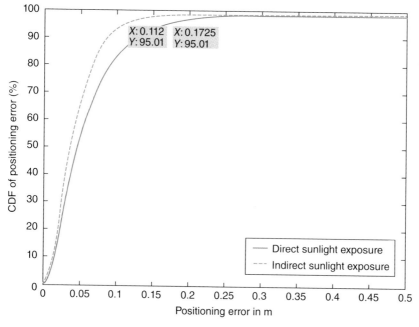

Figure 10.34 CDF curves of positioning error under direct and indirect sunlight exposure assuming wrong positions of LED bulbs and orientation angle of the receiver.

as an acceptable service coverage rate, the proposed system will be able to deliver an accuracy of 17.25 cm (11.2 cm), within indoor environments with direct (indirect) sunlight exposure.

10.4   Conclusions

In this chapter, we have investigated several fundamental research areas of indoor positioning using VLC technology based on LEDs. A survey of different positioning techniques is discussed in detail comparing their benefits and problems. The use of BFSA protocol is addressed as a solution to the channel multi-access problem. The feasibility of this protocol is then shown, considering realistic modulation bandwidth of a typical LED. After noise analysis, simulation results for indoor environments both with direct and indirect sunlight exposure are described. Results indicate that the proposed system is able to provide indoor positioning service with a precision of 95% within 17.25 cm, assuming direct sunlight exposure, and a precision of 95% within 11.2 cm when indirect sunlight exposure is assumed, taking possible wrong positions of LED bulbs and orientation angle of the receiver into consideration.

References

[1] M. Kavehrad and W. Zhang, *Light Positioning System (LPS), Visible Light Communication*, Cambridge University Press, to be published.

[2] Y. U. Lee and M. Kavehrad, "Two hybrid positioning system design techniques with lighting LEDs and ad-hoc wireless network," *IEEE Transactions on Consumer Electronics*, vol. 58, no. 4, pp. 1176–1184, 2012.

[3] Y. U. Lee, S. Baang, J. Park, Z. Zhou, and M. Kavehrad, "Hybrid positioning with lighting LEDs and Zigbee multihop wireless network," in Proceedings of the SPIE 8282, Broadband Access Communication Technologies VI, pp. 82820L–82820L-7, 2012, San Francisco, CA, January, 2012.

[4] Federal Communications Commission, "Location-Based Services: An Overview of Opportunities and Other Considerations," [Available Online]: http://transition.fcc.gov/Daily_Releases/Daily_Business/2012/db0530/DOC-314283A1.pdf (accessed April 18, 2015).

[5] M. Kavehrad, "Sustainable energy-efficient wireless applications using light," *IEEE Communications Magazine*, vol. 48, no. 12, pp. 66–73, 2010.

[6] S. Bhardwaj, T. Ozcelebi, R. Verhoeven, and J. Lukkien, "Smart indoor solid state lighting based on a novel illumination model and implementation," *IEEE Transactions on Consumer Electronics*, vol. 57, no. 4, pp. 1612–1621, 2011.

[7] M. Kavehrad, "Optical wireless applications: a solution to ease the wireless airwaves spectrum crunch," *Proceedings-SPIE OPTO. International Society for Optics and Photonics*, vol. 8645, pp. 86450G–86450G, 2013.

[8] C. Wang, C. Huang, Y. Chen, and L. Zheng, "An implementation of positioning system in indoor environment based on active RFID," in Proceedings of the 2009 Joint Conference Pervasive Computing, pp. 71–76, Taipei, Taiwan, December, 2009.

[9] V. Otsason, A. Varshavsky, A. LaMarca, and E. de Lara, "Accurate GSM indoor localization," in Michael Beigl, Stephen Intille, Jun Rekimoto, Hideyuki Tokuda (eds), *UbiComp 2005, Lecture Notes Computer Science*, Vol. 3660. Springer-Varlag, Berlin Heidelberg, 2005, pp. 141–158.

[10] K. Kitamura and Y. Sanada, "Experimental examination of a UWB positioning system with high speed comparators," in Proceedings of the IEEE International Conference on Ultra-Wideband, pp. 927–932, Singapore, September, 2007.

[11] Y. Liu and Y. Wang, "A novel positioning method for WLAN based on propagation modeling," in Proceedings of the 2010 IEEE International Conference on Progress in Informatics and Computing, pp. 397–401, 2010.

[12] N. S. Correal, S. Kyperountas, Q. Shi, and M. Welborn, "An UWB relative location system," in 2003 IEEE Conference on Ultra Wideband Systems and Technologies, pp. 394–397, Reston, VA, November, 2003.

[13] L. Taponecco, A. A. D'Amico, and U. Mengali, "Joint TOA and AOA estimation for UWB localization applications," *IEEE Transactions on Wireless Communications*, vol. 10, no. 7, pp. 2207–2217, 2011.

[14] H. Liu, H. Darabi, P. Banerjee, and J. Liu, "Survey of wireless indoor positioning techniques and systems," *IEEE Transactions on Systems, Man, and Cybernetics, Part C (Applications and Reviews)*, vol. 37, no. 6, pp. 1067–1080, 2007.

[15] W. Zhang and M. Kavehrad, "Comparison of VLC based indoor positioning techniques," in Proceedings of the SPIE 8645, Broadband Access Communication Technologies VII, pp. 86450M, 2013, San Francisco, CA, January, 2013.

[16] T. Q. Wang, Y. A. Sekercioglu, A. Neild, and J. Armstrong, "Position accuracy of time-of-arrival based ranging using visible light with application in indoor localization systems," *Journal of Lightwave Technology*, vol. 31, no. 20, pp. 3302–3308, 2013.

[17] Z. Zhou, M. Kavehrad, and P. Deng, "Indoor positioning algorithm using light-emitting diode visible light communications," *Journal of Optical Engineering*, vol. 51, no. 8, pp. 085009-1–085009-6, 2012.

[18] Y. Kim, J. Hwang, J. Lee, and M. Yoo, "Position estimation algorithm based on tracking of received light intensity for indoor visible light communication systems," in 2011 IEEE 3rd International Conference on Ubiquitous and Future Networks, pp. 131–134, 2011, Dalian, China, June 15–17, 2011.

[19] A. Küpper, *Location-Based Services: Fundamentals and Operation*. John Wiley & Sons, Ltd, Chichester, 2005.

[20] A. Kushki, K. N. Plataniotis, and A. N. Venetsanopoulos, *WLAN Positioning Systems: Principles and Applications in Location-Based Services*. Cambridge University Press, Cambridge, 2012.

[21] K. Panta and J. Armstrong, "Indoor localisation using white LEDs," *Electronics Letters*, vol. 48, no. 4, pp. 228–230, 2012.

[22] S. Jung, S. Hann, and C. Park, "TDOA-based optical wireless indoor localization using LED ceiling lamps," *IEEE Transactions on Consumer Electronics*, vol. 57, no. 4, pp. 1592–1597, 2011.

[23] Wikipedia, [Available Online]: http://en.wikipedia.org/wiki/Photogrammetry (accessed April 18, 2015).

[24] G. Yun and M. Kavehrad, "Indoor infrared wireless communications using spot diffusing and fly-eye receivers," *Canadian Journal of Electrical and Computer Engineering*, vol. 18, no. 4, pp. 151–157, 1993.

[25] G. Yun and M. Kavehrad, "Spot-diffusing and fly-eye receivers for indoor infrared wireless communications," in 1992 IEEE International Conference on Selected Topics in Wireless Communications, 1992. Conference Proceedings, pp. 262–265, 1992, Vancouver, BC, June 25–26, 1992.

[26] T. Tanaka and S. Haruyama, "New position detection method using image sensor and visible light LEDs," in 2009 IEEE 2nd International Conference on Machine Vision, pp. 150–153, 2009, Dubai, December 28–30, 2009.

[27] G. B. Prince and T. D. C. Little, "A two phase hybrid RSS/AoA algorithm for indoor device localization using visible light," in 2012 IEEE Global Communication Conference, pp.3347–3352, 2012, Anaheim, CA, December 3–7, 2012.

[28] S. Y. Jung, S. Hann, S. Park, and C. S. Park, "Optical wireless indoor positioning system using light emitting diode ceiling lights," *Microwave and Optical Technology Letters*, vol. 54, no. 7, pp. 1622–1626, 2012.

[29] Wikipedia, [Available Online]: http://en.wikipedia.org/wiki/Global_Positioning_System (accessed April 18, 2015).

[30] N. Abramson, "THE ALOHA SYSTEM: another alternative for computer communications," in Proceedings of the November 17–19, 1970, Fall Joint Computer Conference, pp. 281–285, 1970, Houston, TX, November, 1970.

[31] N. Abramson, "The alohanet-surfing for wireless data," *IEEE Communications Magazine*, vol. 47, no. 12, pp. 21–25, 2009.

[32] N. Abramson, "The throughput of packet broadcasting channels," *IEEE Transactions on Communications*, vol. 25, no. 1, pp. 117–128, 1977.

[33] L. G. Roberts, "ALOHA packet system with and without slots and capture," *ACM SIGCOMM Computer Communication Review*, vol. 5, no. 2, pp. 28–42, 1975.

[34] S. A. Ahson and M. Ilyas, *RFID Handbook: Applications, Technology, Security, and Privacy*. CRC Press, Boca Raton, FL, 2010.

[35] J. M. Kahn and J. R. Barry, "Wireless infrared communications," *Proceedings of the IEEE*, vol. 85, no. 2, pp. 265–298, 1997.

[36] W. Welford and R. Winston, *High Collection Nonimaging Optics*. Academic Press, San Diego, CA, 1989, pp. 53–273.

[37] K. P. Ho and J. M. Kahn, "Compound parabolic concentrators for narrowband wireless infrared receivers," *Optical Engineering*, vol. 34, no. 5, pp. 1385–1395, 1995.

[38] X. Ning, W. Roland, and J. O'Gallagher, "Dielectric totally internally reflecting concentrators," *Applied Optics*, vol. 26, no. 2, pp. 300–305, 1987.

[39] J. Wei, Z. Haitao, Y. P. Gong Mali, Y. Xin, Z. Kai, and J. Feng, "Nonimaging concentrators in optical wireless communications," *Laser Technology*, vol. 27, no. 4, pp. 311–316, 2003.

[40] G. Archenhold, "Health and safety of artificial lighting," *Mondo Arc*, vol. 63, pp. 111–118, 2011.

[41] W. Zhang and M. Kavehrad, "A 2-D indoor localization system based on visible light LED," in IEEE Photonics Society Summer Topical Conference—Optical Wireless Systems Applications, pp. 80–81, 2012, Seattle, WA, July 9–11, 2012.

[42] W. Zhang, M. I. S. Chowdhury, and M. Kavehrad, "Asynchronous indoor positioning system based on visible light communications," *Optical Engineering*, vol. 53, no. 4, pp. 045105-1–45105-9, 2014.

[43] L. Lin and H. C. So, "New constrained least squares approach for range-based positioning," in Proceedings of the of European Signal Processing Conference, pp. 1999–2003, 2011, Barcelona, Spain, August 29 – September 2, 2011.

[44] T. Komine and M. Nakagawa, "Fundamental analysis for visible-light communication system using LED lights," *IEEE Transactions on Consumer Electronics*, vol. 50, no. 1, pp. 100–107, 2004.

[45] A. J. C. Moreira, R. T. Valadas, and A. M. De Oliveira Duarte, "Optical interference produced by artificial light," *Wireless Network*, vol. 3, no. 2, pp. 131–140, 1997.

[46] R. G. Smith and S. D. Personick, "Receiver design for optical fiber communication systems," in *Semiconductor Devices for Optical Communication, Topics in Applied Physics*, H. Kressel and G. Arnold, eds, vol. 39. Springer DE, Heidelberg, Germany, 1982, pp. 89–160.

[47] A. P. Tang, J. M. Kahn, and K. Ho, "Wireless infrared communication links using multi-beam transmitters and imaging receivers," in Conference Records of the 1996 IEEE Interenational Conference on Communications, pp. 180–186, 1996, Dallas, TX, June 23–27, 1996.

Index

Short-Range Optical Wireless: Theory and Applications, First Edition. Mohsen Kavehrad, M. I. Sakib Chowdhury and Zhou Zhou.
© 2016 John Wiley & Sons, Ltd. Published 2016 by John Wiley & Sons, Ltd.